THE UNITED STATES AFTER WAR

THE UNITED STATES AFTER WAR && *THE CORNELL*

UNIVERSITY SUMMER SESSION LECTURES, WITH

AN INTRODUCTION BY C. W. DE KIEWIET

ALVIN H. HANSEN

F. F. HILL

LOUIS HOLLANDER

WALTER D. FULLER

HERBERT W. BRIGGS

GEORGE D. STODDARD

Essay Index Reprint Series

BOOKS FOR LIBRARIES PRESS
FREEPORT, NEW YORK

STANDARD BOOK NUMBER:
8369-1069-9

LIBRARY OF CONGRESS CATALOG CARD NUMBER:
69-18571

PRINTED IN THE UNITED STATES OF AMERICA

CONTENTS

*

INTRODUCTION

BY C. W. DE KIEWIET

*Dean of the College of Arts and Sciences,
Cornell University*

SOCIAL PLANNING FOR TOMORROW

BY ALVIN H. HANSEN

*Professor of Political Economy,
Harvard University*

AGRICULTURE IN THE POSTWAR ECONOMY

BY F. F. HILL

*Professor of Land Economics,
Cornell University*

CONTENTS

INTRODUCTION

*

BY C. W. DE KIEWIET

*DEAN OF THE COLLEGE OF ARTS
AND SCIENCES, CORNELL UNIVERSITY*

WHEN the first World War ended the crowds sang and danced in the streets of the victorious capitals. At the end of a second and greater World War, marked by a still completer victory, rejoicing is tempered. In most of the lectures in this volume, there is a note of modesty, occasionally of anxiety. In spite of the magnitude of victory, men contemplate the future with a troubled gaze.

In May, 1918, defeat at the hands of the Central Powers was still a possibility. Only in August or September had the victory of the Allies become certain. The swift movement from fear to relief explains the dancing and singing; it also explains the unpreparedness with which the nations then faced the problems of peace. In the second World War the United Nations were certain of victory more than a year before Germany and Japan actually collapsed. When victory came, the names of Dumbarton Oaks, Bretton Woods, and San Francisco marked a careful and deliberate effort to avoid the incoherence, the obliquity, and the haste of the meetings in Versailles in 1919.

In 1945, peace does not simply mean a blessed escape from fear; it means a clearer knowledge than ever the men of 1918 had of the great burdens that have to be shouldered in our domestic and international life.

At the end of the first World War the conviction was

widely held in England and the United States that victory meant a return to the conditions that had prevailed before 1914. It was believed that the economic dislocations of the war were temporary; its damage could be repaired. There is no such conviction now. If it is correct to say that the first World War was fought to preserve the *status quo*, it is also correct to say that the second World War was fought for the privilege of giving shape to a new order.

Looking back upon the events that brought us to war with Germany and Japan, we can see that America had thrown down the gauntlet before the military dictatorships long before Lend-Lease or Pearl Harbor. At the very time that Nazi Germany entered upon its own totalitarian experiment, and at approximately the same time that Japan began to create a new Empire in the Far East, the President and Congress of the United States began the most ambitious program of social and economic legislation in many generations of American history. The depression had, in effect, compelled the greater nations to find an answer to the question: What is the most valid use which a modern nation can make of its material resources, of its technical equipment, of its industrial capacity? The German and Japanese answer was to create war machines pointed toward plunder and conquest. The American answer was consistent with the broad trends of American history itself. It is not necessary to agree with the detailed legislation of the New Deal to recognize that its purpose was to increase the sum of human welfare. New Deal legislation at least put butter before guns, child welfare before battleships, labor conditions before military power, soil conservation before foreign conquest. Between the

4

democratic and fascist answers to the great economic depression was an incompatibility that in the long run meant the impact of war itself.

To read these lectures is to recognize more clearly than ever that the greatest benefit of victory is the resumption of the effort to define the purposes of our social, our economic, and our political life. Before us lies the task of a comprehensive reformulation of the relations within society and between societies.

A principal function of a great university is to prepare the ground for the decisions which a society must take in order to live and grow. While a university is not itself a center of decision, it is quite properly the center of discussion. The freest discussion and examination are its privilege and duty. To decide upon a course of action is the responsibility of politicians, for they must turn thought and knowledge into deeds. The man in the street is often impatient with the wordiness of legislatures, perhaps because he instinctively knows that maundering and meandering debate reveals an inability to take the road that leads from competent knowledge to enlightened action. Yet debate itself is a vitally practical function, provided that its purpose is to examine alternatives, to clarify issues, and finally to prepare the way for action and decision. In a lecture series which most opportunely ended in the week of peace itself, Cornell University has made a contribution to the debate necessary to restore the world creatively to the conditions of peace.

The war has destroyed some of the world's centers of science and thought; it has crippled others. Political institutions have been broken and bankrupted. The centers to which the world can look for physical help, for political guidance, and for new thought have been reduced to a few nations, less than the proverbial handful. It cannot be immodest to say that, as the war ends, America holds a place of frightening responsibility. We should recognize that the floodlights are playing upon us, upon our return to a peacetime economy, and upon our efforts to meet urgent issues of domestic and international society. The crowd of foreign students in American universities will be far greater than before the war. But American influence will go further. The laws we pass, the strikes we settle or fail to settle, the degree of free economic enterprise we maintain or of economic planning we undertake, the depressions we undergo or the prosperity we create—these will be the material of study in many places. What the next decade does in America will influence the shape of things in the world as a whole.

The world is caught in a flood tide of change for which the war is only partly responsible. In another generation we shall perhaps recognize that the reconstruction of nations like Holland, Italy, or France was as nothing compared with the transformation undertaken by India and China. These great populations are committed, no matter what the travail, to acquire the technical equipment of modern society. But their technological transformation, especially if it be aggressively undertaken, will provoke the social and economic problems of industrialized societies. No more need be said to make it clear that the modern world is interested in more than America's technological leadership. It is concerned also with the resourcefulness which America may develop in meeting upon its own soil the social and economic issues of the age.

The reader will quickly notice that none of the lecturers falls into the trap into which so many people fell at the end of the first World War. Nobody contemplates a happy *status quo* to which we are easily to return. It is true that this generation has paid the full price in fear, blood, and discomfort for a new life freer of insecurity and unrest. These lectures are a small part of the evidence that we are little likely to delude ourselves that there is an easy entry into a relaxed and happy new age. Beneath confident words there are undertones of grimness and an anticipation of work, struggle, and anxiety. Professor Briggs, for example, makes it excellently clear that the new institutions created by our diplomacy are merely tools with which statesmen must work conscientiously in order to profit their own countries and the world. His explanation that diplomacy is still confused by the contradictory propositions that nations are equal in their sovereignty and unequal in their power is stressed. It is a warning against loose optimism about the new organs of world security.

Between the lecturers there is both agreement and disagreement. Sometimes the disagreement is sharp. The stuff of contention separates Mr. Fuller, for example, from Mr. Hollander and Mr. Hansen, and possibly Mr. Hollander from Mr. Hill. Running through the chapters of these four lectures in particular are two main lines of reasoning. The first is aggressively followed by Mr. Fuller. In our society the constructive and responsible leader is the businessman. Business is youthful and therefore always creative. Although it measures its success by material profit, it still follows the Golden Rule, yielding to justice and compassion. Business is society freely at work for its chosen ends. It is therefore the most democratic of all activities. Where business enterprise is free, society itself is free. "The basic economic liberty, the one which needs to be most closely guarded, is freedom of opportunity." Mr. Fuller is confident that business has the initiative, the skill, and the wisdom, under favorable circumstances, to move society forward toward welfare and prosperity.

The second line of reasoning is that modern industrial society is growing steadily less democratic. Decisions of vital importance to millions of men and women are reached in places beyond popular control. An active democracy therefore demands that these places be brought under greater political control, so that society shall have more influence over the decisions which affect its welfare. The economic environment of the great mass of Americans is the factory, the industrial corporation, the civil services, and the public utilities. To the great mass of Americans, in the words of Mr. Hansen, "economic opportunity means largely a chance to get a job." Opportunity leads not to fortune or high office, but to jobs, steady wages, and security. For that reason, laws should be written for the masses rather than for individuals. Mr. Hollander would probably translate Mr. Hansen's thought into the statement that a trade union has a higher claim than a business executive. Not free enterprise for individuals but planning for the masses is the first need of the postwar world.

Postwar planning? By all means. Mr. Briggs and Mr. Stoddard would probably agree with their four colleagues.

The one obviously favors planning in international relations, and the other in educational policies.

Planning for what? The answer still reveals a meeting of minds. Mr. Hill's phrase, "reasonably full and continued employment," is the key to the most essential agreement between himself and the other lecturers. Mr. Fuller is a conservative in his economic thought; Mr. Hollander is more radical. But whatever the point of view, there is real agreement upon the need of security for the American worker, real agreement that labor is not simply a commodity but is an end in itself. The four freedoms are the proper context for labor on the farm, in the factory, and in the office. Let there be free enterprise, and let its rewards be stimulating, but let it be agreed that labor in the twentieth century is no longer a mere tool of production with a price regulated by the market. The welfare of labor itself has become a primary goal of national life. These sentences are the bridges which connect each lecture with its neighbor's territory.

Planning by whom? We now leave the zone of agreement. Mr. Fuller's answer is not Mr. Hollander's answer, and still less Mr. Hansen's or Mr. Hill's answer. "Prosperity cannot be handed down from Washington or from the Statehouse," asserts Mr. Fuller. Mr. Hollander is very certain that "industry, labor, and agriculture, working and planning together" will still require "the full assistance of government" to do the job. Mr. Hill makes it very plain how much American agriculture already exists within a structure built by laws and governmental agencies, and in several places suggests reasons why agriculture may become even more than ever the concern of governmental enterprise. There is danger that returned veterans may buy farm land in an inflated market and then operate it, to their serious disadvantage, in a period of falling commodity prices. It is not very venturesome to predict that the veterans will be of the greatest concern to the national government for a generation to come. Many an American soldier has spent desolate hours in a foxhole daydreaming about chicken farms or ranches. To what extent is society prepared to leave such men exposed to the rigors of adverse market prices?

This war has been a war of planning. The transport of goods over the Atlantic alone was a triumph of co-ordination of which the story will never be fully told, for it was a vast concentrated effort of shipyards and industrial plants, of ships upon the sea and planes in the air, of convoys and escorts, of stevedoring and warehousing. The war ended with a startling demonstration of planned effort in the creation of the atomic bomb. In the first shocked excitement of the detonation at Hiroshima, it was of the marvels of modern science that men spoke. Science had indeed given us a date to set beside the greatest dates of scientific history. But was there a single name to write on the roll of Galileo and Copernicus? The answer is that there was not one name, nor several names, but indeed too many names to be so written. For the unleashing of atomic power was the fruit of an expenditure great enough to affect the national debt, of a teamwork of scientists that took high talent from the universities of Great Britain, Canada, and the United States, of great mining excavations of radioactive earths, of engineering projects that turned villages into towns, of the employment of enough men to fill several army divisions. By planning on a courageous scale, science was hastened along its road to a conclusion that it would otherwise have reached who knows how much more slowly.

The mind of the attentive reader will from time to time pause at comments that are especially provocative. There is Mr. Stoddard's arresting statement that in our educational planning a good kindergarten is as important as a college, and the revelation of some of the serious gaps in America's educational structure. Mr. Hill makes an essential comment in his concluding paragraphs. He is afraid that modern society may be setting restraints upon its own productivity. A dozen years ago we plowed under cotton and corn; at the close of the war there are demands that we "plow under" surplus industrial products. Trade unions, cartels, and political pressure groups have discovered the use of restrictive measures. There is nowadays much skill in producing work stoppages, in limiting output, in blocking the efficient movement of goods. Is modern society becoming more or less laborious? Western society originally became great, setting its impress upon all the world and its peoples, because the principle of work achieved an authority and a dignity which it had nowhere else. In an age when concern with the life to come was paramount, Benedict of Nursia insisted upon work in this world, upon man's capacity to produce and to create. Work produces food and tools. It creates cathedrals and paintings. It establishes schools and nurtures youth. Modern society becomes unsteady and confused unless it has initiative, ambition, enterprise, momentum. There is a logical relationship between full employment and full production. Robert Owen knew this when he wrote more than a hundred years ago that the ordinary labor of individuals judiciously directed to objects of a productive nature would be more than sufficient to maintain them in a high degree of comfort and that human exertion if properly directed would increase the objects of desire almost to infinity.[1]

---

[1] I am indebted for this reference to my student Miss Lois Citrin.

SOCIAL PLANNING
FOR TOMORROW

*

BY ALVIN H. HANSEN

PROFESSOR OF POLITICAL ECONOMY,
HARVARD UNIVERSITY

A FEW WEEKS AGO I participated in a discussion on planning in the postwar world. Among the participants were some who expressed, in general abstract terms, vigorous opposition to planning in all forms. Theoretical arguments, drawn largely from Professor Hayek's book *The Road to Serfdom,* were expounded. There followed an address by a middle-of-the-road Republican Senator, well known for his common-sense approach to practical problems, and utterly devoid of doctrinaire views. "I approach the topic," he said, "from the standpoint of the day-to-day problems that come before me as a member of Congress." Before he was through, he had touched upon innumerable domestic and international problems upon which he was compelled, as a member of Congress, to have an opinion and about which something had to be done. For him, it was not an abstract question about planning or not planning. For him some workable solution had to be found that might prove reasonably satisfactory, else up bobbed these questions again.

The plain fact is that all advanced individual nations have moved very far away from the atomistic individualism of the mid-nineteenth century. Then economic opportunity meant essentially a chance to operate your own farm or small business. Today economic opportunity means largely a chance to get a job. Then the bulk of the population lived in the country—on farms or in small villages. Today they live in great urban centers. Industrialization and urbanization have come upon us with a speed which no one could have imagined in 1850. Torn from the old individualistic pattern of work and living into a society characterized by great factories and giant cities, modern man must erect a new social structure adapted to the changed conditions. In an industrialized and urbanized society, the individual cannot order his life alone. He can meet the problems of living only by joint action with his fellows. This is the great challenge of modern life—how to reconcile the rights and freedoms of the individual in a society in which group and community action is a necessary basis for successful living. This is the problem of democratic planning in the modern world.

Planning is, however, not something new. Its form and content have, it is true, changed. In the early days of our country, planning had to do, so far as economic matters were concerned, with creating an environment in which the citizens could successfully engage in farming or in business as small shopkeepers or tradesmen. It meant essentially a legal and economic environment favorable for the starting of a small enterprise. It became a primary responsibility of government to ensure the right to set yourself up in business. This right was not the product of automatic forces, as some naïvely suppose. This right was fought for in the great social and economic movements of the early nineteenth century. It was fought for in the antimonopoly movements, in the struggle for free banking, in the great Homestead movement which finally won the right to free land. The history of the first half of the nineteenth century was alive and glowing with great human programs—social plans to ensure and maintain economic opportunity.

A program to keep open the door for new enterprise remains also today a vital and necessary part of economic planning. Among other things this means a revitalization of our antitrust laws and their enforcement, reform of our patent system, and an adequately financed Institute for Technical and Scientific Research to aid new and small business.

By and large the right to establish a business or to acquire free land was adequate, in the nineteenth century, to maintain economic opportunity and to ensure the right to "life, liberty, and the pursuit of happiness." This is no longer the case. In all modern countries, the trend of technology, whether in industry, transportation, or distribution, restricts economic opportunity, for the overwhelming majority, to the getting of a job—not to establishing a business of your own. If, therefore, we are to keep open the door of economic opportunity, under modern conditions, it becomes necessary for modern society to undertake, as a primary responsibility, the maintenance at all times of adequate employment opportunities. Just as the right to free land was the watchword of economic opportunity a hundred years ago, so the right to useful, remunerative, and regular employment is the symbol of economic opportunity today. The Murray full employment bill is today's counterpart of the Homestead Act of a century ago. The Homestead movement of the 1840's represented a great struggle for human rights and economic opportunity. It was fought by the forces of reaction. But the issue could not be evaded. So also with the full employment program today. It involves elemental human rights—the right to life, liberty, and the pursuit of happiness. So long as 80 to 90 per cent of the population cannot earn a livelihood except by getting a job, the issue of full employment will not down. Of this we may be certain.

The White Paper on Employment Policy, issued by the British Government over a year ago, is the first formal acceptance by a leading nation of the primary aim and responsibility to maintain a high and stable level of employment. This was followed early this year by a similar declaration by the Canadian Government. In setting as its aim a high and stable level of employment, the Canadian Government specifically stated that it was not selecting a lower target than "full employment." Rather it was mindful that it was breaking new ground and that it needed full public understanding and support to achieve its high goal. A mere declaration will not achieve full employment.

In this lecture, however, I am not prepared to discuss all the innumerable difficulties we shall encounter in the pursuit of a full employment goal. The fact that this experiment will prove difficult will not permit us to escape it. The problems which we shall be compelled to face will command the ingenuity and resourcefulness both of theoretical economists and practical statesmen in many decades to come. What I wish to stress here is my deep conviction that the political democracies of our time must undertake this new responsibility. We have reached a stage in economic evolution in which social planning must go beyond providing the economic opportunity to set up a business. For the great majority of citizens, in the world in which we live, economic opportunity means the right to a job.

The old right and the new one are interrelated. Freedom to choose among job opportunities in a free society presupposes the right and the opportunity for as many citizens as possible to establish a private enterprise. This right and this opportunity is important, not only for those who can and wish to become independent entrepreneurs, but also for those who remain employees. This is true because only in a private enterprise economy can the great mass of wage and salaried workers enjoy the essential freedom of choice of employment among thousands and thousands of different employers. In a totalitarian society there is only one employer—the state. It is a vitally important safeguard for the preservation of personal liberty that the citizens of a free society shall enjoy the opportunity to choose between numerous employers, including private entrepreneurs, cooperative societies, and governments, federal, state, and local.

Thus the great goal of full employment, if it is to be achieved in a free society, involves planning to make the market economy function in a workable manner so as to provide adequate employment opportunities together with the privilege of choice between different employers.

If the democratic countries were not now planning and developing new institutional arrangements designed to make the market economy function more effectively than it did in the past, the future would be black indeed. Those who think that a reversion to the institutional arrangements of the nineteenth century would give us, in the world we live in, stability and prosperity are not realistic.

They are nostalgic dreamers. They are fighting for a lost cause. We cannot meet the problems of today by institutions suitable to conditions that no longer exist. We need, and we are in fact devising, new plans both domestic and international.

The old market economy has broken down. It failed us utterly in the two decades between the two World Wars. In England unemployment, never falling below 10 per cent, reached in some years 22 per cent of the labor force and averaged 14 to 15 per cent for the entire two decades. In the United States, taking account of the whole interwar period, the spotty twenties and the depressed thirties, unemployment averaged about 12 per cent, and in the worst years, 1932–1933, reached 24 to 25 per cent of the labor force.

All hopes for the restoration of the old prewar economy were dashed in the great depression which shook the entire world and fanned the flames of the ensuing terrible world conflagration. International economic co-operation was completely cast aside. Economic warfare became the rule. The old order was destroyed. A new economic structure can be erected only on the basis of new institutions. We must rebuild the market economy, both in the domestic and the international spheres, so as to prevent similar disastrous breakdowns in the future. We are confronted with the task of devising new machinery, suitable to modern conditions, under which the market economy can operate effectively at high and stable levels of income and employment.

Under the market economy, good employment opportunities depend upon an adequate demand for goods and services. As the British White Paper on Employment Policy put it: "A country will not suffer from mass unemployment as long as the total demand for its goods and services is maintained at a high level." "Demand," in economic terminology does not mean "need" or "desire"; it means "outlay" or "expenditure." Adequate total demand means an adequate total outlay whether by individuals, business, or government for goods and services.

Some critics have had a field day pointing out that high total outlays will not cure pockets of unemployment in stranded areas; nor will it cure seasonal, frictional, and technological unemployment. They have pointed out that expenditures on land, old houses, securities already outstanding, and the like do not create employment. They have moreover noted that increased outlays, if wages are increased more rapidly than increases in productivity, would merely result in high prices without increasing employment.

These criticisms,[1] while accurate enough, are not very

---

[1] Critics have also concentrated their attacks on the phrase "full employment." In this connection I should like to quote the following, which is a footnote taken from my pamphlet *After the War—Full Employment,* published by the National Resources Planning Board in February, 1943, long before the recent controversy developed. The footnote (p. 3) is as follows:

"The term 'full employment' is often misunderstood and requires brief explanation. A little reflection will make it clear that in a highly dynamic society in which new industries are developing and some old ones are declining we must retain a high degree of labor mobility. In like manner, regional population shifts will occur in an expanding, developing economy. In addition, in a democratic society with freedom of occupational choice, some considerable labor turnover is not only inevitable but, indeed, beneficial. Without this, personal freedom could not be maintained. Thus, a dynamic and mobile society requires some considerable shifting of jobs. Accordingly there will always be, in a society such as ours, a large amount

21

brilliant. They are mainly old truths that intelligent people may take for granted without constant reiteration. Neither the British or Canadian Governments, nor the sponsors of the American full employment bill, are naïve enough to overlook these matters. The points raised have been dis-

---

of transitional unemployment. In addition, there will inevitably remain, even under the best planning by business and industry, a considerable amount of seasonal unemployment. For these reasons, in an economy as large as that of the United States, it is probable that at 'full employment' there would be at any one time between 2 to 3 million temporarily unemployed.

"It must, moreover, be recognized that the concept of full employment can acquire a definiteness only when it is conceived within the pattern of social customs and institutional arrangements which determine the size of the labor force and the customary hours of work. Thus, the labor force will be affected by the customary age of retirement and also by established practices with respect to minimum years of schooling and prevailing practices with respect to the age of entrance into industry. With respect to the hours of work, 'full employment' obviously does not mean that the population will work at the maximum possible hours that human endurance makes feasible. On the contrary, the prevailing hours of work will be determined by legislation, collective bargaining, and customary practices. The concept of 'full employment' presupposes that the normal labor force is working at the customary and prevailing work week.

"Finally, the concept of full employment relates only to those members of the community who are 'employable.' Unemployables, whether by reason of physical or mental defects, are not a part, properly speaking, of the labor force. The problem of what to do with these elements of our population is certainly an important one to which a democratic society must seriously address itself, but it is not part of the problem of achieving 'full employment.' It is, however, emphatically true that if we achieve a 'full employment' economy, the relative scarcity of labor thereby created will force our society to tackle more vigorously the problem of training and educating some portion of the 'so-called' unemployables, making them sufficiently efficient to be added to the employable labor force. This we have never done in the past because there has typically been available a reservoir of unemployed to draw upon. A full employment society, continuously maintained, will discover that it is quite possible through education and training to reduce very substantially the proportion of the population which has in the past been regarded as unemployable."

cussed, both in the theoretical and the applied literature, again and again. What needs *now* to be stressed is the all-important fact that the market economy cannot function even tolerably well unless the total outlay on goods and services, public and private, is maintained at a high and stable level. This is the central problem to which we must address ourselves, if we really want the free market economy to work. How can we ensure that the total outlay on goods and services will be adequate to provide continuing full employment?

The outlays on goods and services, as elaborated in nearly all current full employment documents, can conveniently be divided into four categories: (1) private consumption expenditures, (2) private capital outlays, (3) current services of government, and (4) government capital outlays on public works and developmental projects.

The first of these—private consumption expenditures—can be fairly closely estimated for different levels of employment. We know from all past experience that private consumer expenditures cannot be expected to reach a level adequate to employ the entire labor force. A rich society such as ours does not and will not spend all its income on consumer goods. We need in the postwar period, it is estimated, a total outlay in goods and services of around 170 billion dollars to provide full employment. But we know from long-established patterns of spending and saving (taking account of probable postwar taxation) approximately what the maximum contribution is which consumers can be expected to make toward that needed total outlay of 170 billion dollars. The probable maximum is not far from 115 billion dollars. This means that around 55 billion dollars must be expended by private business on capital outlays and by federal, state, and local governments on current services or construction projects. Only a drastic change in the expected peacetime tax structure or really fundamental changes in social security and income distribution could alter these figures substantially.

Now while the maximum contribution to total outlay from consumer expenditures can be estimated with fair accuracy, we know very little about what may be expected from year to year in private capital outlays. In some years private investment is high and in some low. It was seventeen billion dollars in 1929, and it fell to only two billion dollars in 1932. Herein lies the essential explanation of the great depression. It is just this utter undependability of private capital outlays that makes the economic system so unstable.

The third category—public outlays in current services—can also be set down as a pretty definite figure. We know what the yearly outlays are on education, police protection, and the usual current government services. From the long-run standpoint we may wish to raise the figure gradually, but from the short-run it does not and should not fluctuate materially. We know pretty accurately therefore what the contribution will be to total demand from this third category of outlay or expenditure.

The fourth category—public capital outlays on public improvement and development projects—like the second is subject to wide variation. Government construction projects for the most part do not have to be made in any one year. We can plan a fifty-billion-dollar program, more or less, according to our prospective needs, over a six- to eight-year period, and we can vary the annual outlays according to the requirements of stability and full employment. If private capital outlays decline, public capital outlays can be stepped up. Thus the public sector can act as a balance wheel to the private sector. In this matter, with adequate planning we could stabilize the construction industry as a whole, taking account both of the public and the private sectors. Public projects should be built in the usual case under private contract. Thus contractors would switch from private to public projects as private capital outlays declined and public outlays took up the slack. But the construction industry, privately owned and operated, would find a stabilized volume of outlays, public and private combined.

Stabilization of the construction industry would go very far toward stabilizing the economy as a whole. But it cannot do the job alone. A broad and comprehensive system of social security and social welfare, combined with a progressive tax structure, acts steadily and continuously as a powerful stabilizing factor. It puts a floor under depression. It acts as a great irrigation system, distributing purchasing power widely over the entire country. Cyclical variation in the standard income tax rate, now that we have current collection at the source, can I believe serve as a useful and effective supplementary measure. But the two main measures upon which we must rely for stability and continuing full employment are (1) a comprehensive and flexible program of public improvement and development projects, and (2) a comprehensive system of social security.

Once these measures are adequately implemented to

sustain and advance the level of income and employment, it will I think be discovered that the business cycle will become something very different from that experienced in the past. The cumulative features which have characterized the cycle for a hundred years would tend to disappear. Under the automatic forces which controlled the cycle in the past, once the downward movement got started, the cumulative process fed on itself. Unemployment spread fear among consumers and reduced the volume of expenditures. Falling prices and falling markets induced pessimism among businessmen and cut off new capital outlays. In contrast, a sustaining social security and developmental program will tend to stop this cumulative process. Thus the cycle, within the framework of an adequate compensatory, developmental, and social security system shorn of its worst cumulative features, may become manageable and susceptible to social control.

Modern governments are just at the threshold of this great experiment. We are still in the kindergarten stage. The stabilization of the construction industry alone involves an immense amount of physical and fiscal planning. It involves city planning and programs of urban redevelopment. It involves a comprehensive housing program, including not only a twenty-year plan for the demolition of substandard houses, but also a long-range program of new residential construction, public and private. It involves a national plan for regional resource development in every part of the country, taking account not only of the great river basins such as the Tennessee Valley, the Columbia, the Missouri, and the Arkansas, but of land and water resources up and down the country which need reclamation, development, and conservation. It involves a thorough modernization of our entire transportation facilities—roadways, airways, waterways, and railways. In these three great fields—urban redevelopment and housing, regional resource development, and transportation—public investment must play an important role if we are to rebuild America on lines commensurate with the potentialities of modern science and modern technology. As we look at the deplorable physical condition of our great cities, the substandard housing both urban and rural, the congested urban transportation facilities, and the wastage of natural resources, it becomes abundantly evident that our greatest deficiencies are precisely in those areas which require large public investment outlays. We need to undertake a great national program of development. And such development would open up new rich fields for private investment.

I have underscored the role of a comprehensive system of social security and social welfare in a program of stability and expansion. Education, health, and nutrition, recreational facilities, and community cultural activities are fundamental in a broad national development program. Of what good are mere brick and mortar if we neglect to develop a healthy, trained, educated, and socially minded citizenry? We are seriously deficient in our great country, not only in terms of natural resource development, but also in terms of human development. Forty per cent of our children grow up in areas deplorably deficient in educational facilities. A disquieting percentage of the young men drafted into the service were adjudged "functional illiterates" or were physically unfit for military duty. These are areas we must not neglect when we plan a well-balanced program of national development and public investment.

Planning for the future, however, cannot stop short at our national boundaries. National planning for stability and expansion involves not only domestic but also international plans. It is to the credit of the leading governments of our generation that we have not drifted on to the end of the war without a program. We are on our way to building a set of new international institutions that fit the needs of the world we live in. We have held, while the war was still being fought, a whole series of international conferences—Atlantic City on Relief and Rehabilitation, Hot Springs on Food and Agriculture, Bretton Woods to devise a new international monetary system and to provide capital for international development, Dumbarton Oaks and San Francisco to give us a charter for a World Government.

One of the main pillars of the United Nations Charter is the Social and Economic Council. Through this council the member nations will dedicate themselves to the continuing task of solving their common economic problems and of achieving international stability and expansion. This requires first and foremost full employment and economic stability within each country. Any country which fails at home cannot be a good neighbor in the family of nations. This is especially true of the great countries. There is nothing that the United States can do which will contribute more to international stability than to achieve a high and stable level of prosperity at home. Yet no country, at least none of the free countries which operate on the basis of a market economy, is immune to economic disturbance from the outside. Depression spreads, we have learned in the interwar years, with devastating effect from country to country. Collective action by the whole family of nations thus becomes necessary. The Social and Economic Council can play an important role toward achieving a workable international order.

The agreements made by forty-four countries at Bretton Woods, together with the inclusion of the Social and Economic Council in the San Francisco Charter, are indications that there is overwhelming agreement throughout the modern world that new international institutions are necessary if we would escape a repetition of the disastrous experiences between the two World Wars. We are determined not to let things drift again. We mean to become masters of our fate. We shall not achieve a hundred per cent perfection. But we do mean to set our standards high. We have become convinced, at long last, that the old machinery will not work. We are no longer afraid to try something new. This is the meaning of the International Monetary Fund, the International Bank for Reconstruction and Development, the International Organization for Food and Agriculture, the Social and Economic Council of the United Nations Charter; and with respect to domestic policy, this is the meaning of such documents as the British White Paper on Employment Policy, the Canadian Paper on Employment and Income, and the Murray full employment bill.

The catalog of programs, which I have just listed, is impressive. I have discussed the international aspects of this program of planning for the future in some detail in my recent book, *America's Role in the World Economy.* I cannot particularize here. The list of institutions which are there described and which are on the way to realization present a strong contrast to the confusion and frustration which characterized economic policy all over the world twenty-five years ago. We have learned, I am convinced, a great deal from the terrible experiences of the last two decades. We are reaching in all the advanced countries an amazing degree of agreement about what we need to do to reshape our world in order to make it again a functioning and manageable system.

I have had in mind, in what I have said, chiefly the problems confronting the free societies—the countries where economic life is ordered mainly on the basis of private enterprise, but with the state playing nonetheless a large and increasing role. It has been aptly called a "mixed system." It is no longer the old simon-pure private enterprise economy. But its most characteristic feature continues nevertheless to be *the market* or *the price system*. This basic characteristic is common to all the free societies—to the United States, Great Britain, the Scandinavian countries, Holland, Belgium, France, Canada, Australia, and New Zealand. It is upon these countries, in particular, that the task devolves to rebuild by means of new institutions, appropriate to modern conditions, a workable world.

Such a world must combine security with progress. There are those who have sought to show that these two goals are in conflict. But I think it can convincingly be shown that this is a highly superficial view. The modern urbanized world, highly interdependent geographically and occupationally, is extraordinarily sensitive to instability. The modern social structure cannot survive the kind of economic instability, domestic and international, we have suffered in our generation. Amidst such chaotic upheavals, progress cannot flourish. And conversely it is not possible to achieve a high degree of stability except in an expanding world.

The restrictive policies of trade unions are most evident in the highly unstable building industry. Instability promotes restriction and contraction. Foreign trade restrictions—import quotas, exchange control, etc.—multiply in periods of deep depression and mass unemployment. Restrictive practices of all kinds are the instinctive defensive mechanisms of a contracting market. Expansion makes it possible to achieve stability without making it necessary to resort to the contrived and artificial stability of restrictive practices.

If we hope to win social and economic stability, we must make sure that productivity and standards of living are continually on the increase. Large-scale governmental support for scientific and technical research deserves to be put high on the agenda of public investment projects. The development of new products, new methods, new industries is an essential condition, not only of expansion and full employment, but also of economic stability and social security. A stagnant society, incapable of raising the standards of living of its people, will not be a secure or stable society.

And now a final word about "planning" and the so-called "road to serfdom." There are those who allege, led by Professors Mises and Hayek, that conscious planning is not compatible with personal freedom. They would rely upon the automatic forces alone. Only such institutions as are

31

basic to the necessary automatic processes are acceptable to them. All other planning is regarded as "bad planning." That at least is the logic of the case, and it is in fact the position of Mises. Hayek, a little less rigid and doctrinaire in his thinking than Mises, is forced to abandon logical consistency in the face of the hard realities of the modern world. Thus he admits, somewhat reluctantly, the necessity of social security and similar measures of social planning.

Hayek fears a world which requires human management. He has a great deal to say about the "rule of law." Yet he confuses automatism in social life with the "rule of law." He forgets that the very concept of the "rule of law" was developed in England just at the time when there was a vast amount of governmental control of economic life. It was precisely because this was the case that the "rule of law" concept was developed. The "rule of law" does not mean that human management is replaced by automatic forces. It means that the conscious management of social life is conducted under established canons which preclude *arbitrary* action. The "rule of law" substitutes rational principles of management for the arbitrary acts of arbitrary men.

The widespread acceptance of social and economic planning by all modern governments is evidence that the hard experiences of recent decades have driven home the lesson that the functioning of modern economic life cannot be left to automatic forces. But it does not mean that we thereby deliver ourselves up to the arbitrary management of irresponsible men. That is not now, and never has been, the method of political democracy. The plans we are devising, both domestic and international, are not ships lets loose on an uncharted sea with instructions to the captains to steer as they see fit. Rational and democratic planning involves the development of "rules of law" which preclude arbitrary action by those who are chosen to administer the plans. This could be illustrated in any one of the plans, domestic or international, which I have reviewed. For example, while the International Monetary Fund does mean that we have abandoned the moorings of the old international gold standard, it will not leave us in international monetary matters adrift upon a sea of arbitrary decisions by the Governing Board of the Fund. New moorings to take the place of the old gold standard are established. "Rules of law" are set up which constitute a framework within which decisions are made.

In the decades that lie ahead, a major task of economic and social statesmanship must be undertaken precisely in this field. We must evolve "rules of law" under which the social planning of the future can be made a rational and democratic method of managing our social order. Only thus can we achieve the high goals of progress, stability, and full employment combined with the undying human values of personal and individual liberty. Without exception, all the great democracies are today embarked upon programs of social planning. In our own Anglo-Saxon tradition, the concept of social control *under law*, not control by irresponsible and arbitrary men, reaches far back into history. It antedates the age of modern capitalism. It reaches back into the early beginnings of local government. And it was never wholly lost sight of even in the heyday of *laissez faire*. We have a rich heritage of legal and political tradition which gives us high confidence and faith that we can achieve the great social and economic goals we seek without losing our personal freedoms.

AGRICULTURE IN
THE POSTWAR ECONOMY

*

BY F. F. HILL

*PROFESSOR OF LAND ECONOMICS,
CORNELL UNIVERSITY*

IN DISCUSSING agriculture in the postwar economy it must be stressed at the outset that agriculture cannot be divorced from the rest of the economy. What happens within our economy as a whole will largely determine what happens in agriculture. If we succeed in making the transition from war to peace without undue inflation or deflation of prices, and with other conditions favorable to full and continued employment, agriculture will be affected and will react in one way. If, on the other hand, our excessive wartime purchasing power should ultimately result in sharp and substantial inflation in the postwar period, agriculture as well as other parts of our economy will be confronted with an entirely different set of economic conditions and will react in a different way. If we should pursue a policy of deflation in the postwar period, such as England pursued after World War I, we can expect a still different economic situation following the present war with different adjustments in agriculture to meet it.

The question of what is going to happen after World War II is an interesting one and one to which all of us would like to have the answer. Will we have further inflation, and if so, when and how much? Will we be able to make a smooth transition from war to peace without unduly rocking the economic boat, or are we headed for a period of sharply falling prices, unemployment, and all of the difficulties we went through in the early 1920's?

I do not pretend to know the answers to these questions. War sets loose powerful economic forces. The strength of these forces is difficult to judge, much less their net effect at a given time on the economy of a given country. Postwar economic developments in some of the other countries which participated in World War I were fairly similar to our own. In other instances they differed sharply. Defeated Germany went through all-out inflation. England went through a period of moderate inflation followed by a long period of deflation. A number of Continental European countries had substantial inflation, later stabilizing at an inflated price level.

Prices in the United States continued to rise for about a year and a half after the end of World War I, then broke sharply, eventually stabilizing about one-third below the level at the end of the war. This period of relative stability began a little over three years after the end of the war and lasted for seven or eight years. Similar periods of price stability beginning approximately three years after the end of the wars and lasting for seven or eight years occurred in the United States after the Civil War and the War of 1812. Following all three wars prices stabilized at levels considerably below the wartime peaks and in each case the period of stability was followed by renewed declines.[1]

Though price movements following the three previous major wars in which the United States has engaged have much in common, there have also been important differ-

[1] A comparison of the wholesale prices of thirty basic commodities in the United States during and following the War of 1812, the Civil War,

ences. It is dangerous to rely too heavily upon historical comparisons in attempting to forecast developments following World War II. Because economic developments in the United States followed a certain pattern after World War I, it does not follow that the story will be precisely the same after World War II. Even if the broad general pattern should be the same, which is quite possible, the timing and extent of different price movements are likely to be sufficiently different to be of great importance to persons engaged in agriculture, business, and industry. In attempting to forecast what is going to happen, one can only make a careful study of the forces at work and attempt to appraise their probable effect. Appraisals of any kind necessarily involved personal judgment and are subject to the errors inherent in such a process.

I am not going to attempt to make an appraisal of postwar economic prospects tonight. Fortunately for me, the subject assigned does not necessarily call for such an appraisal. In any case, it seems to me that our purpose will be better served if I try to point out the characteristics of agriculture as an industry that are important in determining the way it reacts to changing economic conditions, in the hope that such a discussion will lead to a better understanding of agriculture in the postwar period, come what may in the way of economic developments.

For purposes of illustration, I shall make considerable use of economic developments during and following World

and World War I is given in an article entitled "Prices Following Wars," by F. A. Pearson, W. I. Myers, and J. H. Lorie in the May, 1945, issue of *Farm Economics*, published by the New York State College of Agriculture, Ithaca, New York. This article also gives a short discussion of prices in other countries following World War I.

War I. I wish to stress at the outset, however, that in using World War I as a peg on which to hang this evening's discussion, I am not forecasting that developments following World War II will be the same as those following World War I, but am merely using certain historical facts as a basis for describing the way agriculture as an industry reacts to different economic conditions.

WAYS IN WHICH WAR AFFECTS AGRICULTURE

Agriculture is a raw-material-producing industry. Like other industries producing basic commodities, it is subject to violent fluctuations of prices and net profits. In periods of rising prices, the prices of farm products usually rise faster than the prices of nonfarm products and faster than costs. In periods of falling prices, farm prices usually fall faster than nonfarm prices and faster than costs (figure 1).

During wars there is always an increase in the demand for food and fiber, since both are essential to waging war. During wars, wages and profits in the war industries of a price economy must be allowed to rise, at least in the early stages, to bring about rapidly the necessary change-over from peacetime to war production. Higher wages in war industries, together with the needs of the armed services, tend to bring about a rapid reduction of manpower in agriculture. The needs of war industries for steel and other critical materials reduce the supply of machinery and various production supplies available to farmers.

Increased demand for food and fibers on the one hand, together with limited means of production on the other, plus the general inflation which accompanies the methods we so far have followed in financing wars, give rise to sharp increases in agricultural prices. In the early stages of war, farm prices rise faster than costs, and net profits rise sharply, particularly if, as in World War II, farmers are favored with good weather and resulting heavy production. As the war progresses, costs tend to catch up with

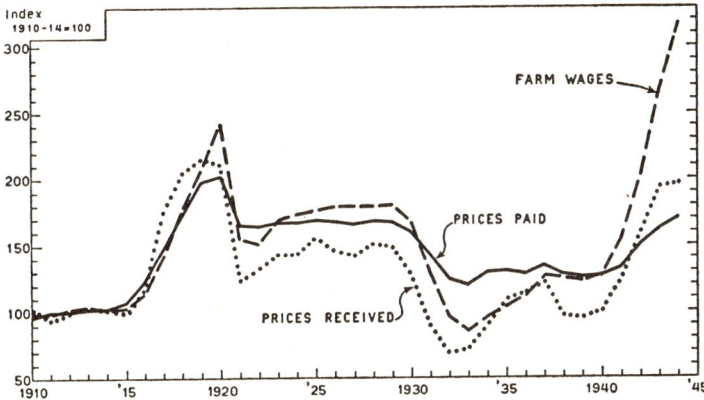

FIGURE 1. INDEX NUMBERS OF UNITED STATES FARM PRICES, FARM WAGES, AND PRICES PAID BY FARMERS FOR ARTICLES PURCHASED, INCLUDING INTEREST AND TAXES, 1910–1944 (1910–14 = 100)

During World War I farm prices rose faster than costs. In 1920 prices began to decline and fell faster and farther than costs. The index of prices received remained below the index of prices paid throughout the twenties and thirties. In World War II prices received have risen faster than prices paid but not so fast as farm wages.

prices, although the relationship is likely to continue fairly favorable throughout the war, which means that farm incomes continue to be good. This type of situation developed in World War I and, so far, history has repeated itself in World War II.

I have pointed out that one of the fundamental characteristics of agriculture is that it is essentially a raw-

41

material-producing industry and, as such, is subject to violent fluctuations of prices and net profits. A second characteristic is that it is an industry with a large number of small businesses in which a part or all of the capital, including land, is supplied by the workers, and in which the home and the business are combined.

When a farmer buys a farm, he buys both a home and a job. It is understandable that he should want to own his own farm, both because he wants to own a home and because he feels that farm ownership is a means of attaining greater job security. Like most of us, farmers make purchases when they have the necessary money, which is usually in periods of rising prices. Since the capital investment in a farm business is relatively large in relation to annual earnings, and since most young men entering farming have relatively little capital, it is a common practice to purchase an equity in a farm, giving a mortgage for the balance of the purchase price, the mortgage to be paid over a period of years out of earnings from the farm business. If farmers had to pay cash for farms, most of them would be old enough to retire before they had enough to buy a farm.

In periods of rising farm prices and incomes, such as we had from 1914 to 1920 and again from 1939 to the present time, many farm tenants, farmers' sons, and farmers who wish to build up operating units of economic size are in the market for farms. The result is that land values rise and farm mortgage indebtedness tends to increase. It has been said that farmers buy farms when they are high and pay for them when the prices of farm products are low. There is a measure of truth in this statement. Farmers act in this fashion, however, not because they fail to realize that land is cheap in depression periods such as the 1930's, but because they do not have the money in such periods to make even a relatively small down payment on a farm. They are buying farms now, even though land values in some areas are reaching high levels, because they have the money with which to buy. Some are paying cash. Others are following the more common practice of purchasing an equity and going into debt for the remainder of the purchase price. Even with rising land values during recent years, it has been possible for farmers to buy farms and either pay for them fully during the war or to build up a sufficient equity so that there is little danger of getting into serious difficulties during the postwar period unless we have a really severe price decline.

The difficulty, of course, is in knowing when we are approaching the peak of prices for farm products. How late in the war period can a farmer buy a farm on a thin equity and still have time enough to build up a safe equity before prices break, if they do break as they have broken after previous wars? That, of course, is the all-important question. In the past, and I assume there will be little difference this time, farmers have continued to bid up the price of land until a break in the prices of farm products has occurred. When prices of farm products and farm land have eventually broken, a substantial percentage of farmers have been left with farms purchased at high prices and heavily mortgaged, which, of course, has led to trouble.

In World War I, average farm real estate values in the United States rose 67 per cent between 1914 and 1920, and in many areas rose more than 100 per cent. Mortgage debt increased by more than 70 per cent between 1914 and 1920, and continued on up until it reached a peak in 1923 nearly 125 per cent above the 1914 level. With the unfavorable price-cost relationships, compared with prewar, that existed in the 1920's, many farmers who were not in debt or who had only modest debts were in difficulties. Farmers who were heavily in debt faced financial disaster.

In World War II, land values started from a lower level than last time, but have risen about the same percentage as during the corresponding period of World War I. During World War II, however, mortgage indebtedness has declined rather than increased. The situation still has many elements of danger, however. Much of the rise in land values and in farm debt resulting from World War I occurred *after* the war was over. If a considerable number of returning veterans and war workers enter the land market after World War II is over, and it seems probable that they will, and if farm prices and profits hold up for a time, as they did after the last war, there is still plenty of opportunity for a boom in farm lands accompanied by increasingly full mortgages to finance purchases. I should not be at all surprised after this war to see the present downward trend in farm mortgage debt, which represents heavy debt liquidation by present owners, reversed and a substantial increase take place. Many farms are now in the hands of older operators who could not afford to retire during the depression of the 1930's and who have been continuing to operate their farms during the war, not only because it was profitable to do so, but also because the normal purchaser of the farm was in the armed services and the nation needed all of the food that could possibly be produced. It is quite possible that many of these farms will change hands after the war, passing from older operators to returning veterans and others at high prices. This possibility creates a real social as well as an economic problem. If farm prices should break and profits decrease as they did after the last war, many a returning veteran may find that he has fought a war only to return to a lifetime struggle of trying to pay for a farm purchased at what may prove to have been too high a price.

One other point should be mentioned in connection with land values and farm debts. Most farms in the United States are owned by individuals. This means that when a farm owner dies the farm is either sold or is passed on to one of his heirs. Quite frequently it is necessary for the heir taking over the farm to mortgage it to obtain funds with which to pay off the other heirs. Transactions of this kind take place daily in periods of high prices and high land values as well as in periods of low prices and low land values. A certain percentage of the mortgages placed on farms during periods of high prices which later lead to difficulty grow out of our particular system of land tenure. The heir who is on the farm, or who wishes to take it over, frequently has the choice of borrowing a substantial amount of money to pay off the other heirs or of forfeiting his opportunity to acquire the farm—perhaps a farm on which he grew up and which he has been farming for a number of years. Practically speaking, he is not in a position to decide when he shall acquire a farm. The decision is more or less forced upon him by circumstances over which he has no control. If he has to assume a heavy debt in connection with the transaction, he may be in difficulty if farm prices and land values subsequently fall.

I should like now to summarize briefly what I have said so far. In periods of rising prices, such as during wars, farm prices rise faster than nonfarm prices and faster than costs. Farm earnings increase and, if the price rise continues for a period of time, large numbers of buyers enter the land market. This is not only because of the opportunity to make wartime profits from farming, but because farming is a business under our traditional system of land tenure in which the operator seeks to own the means of production, including land, and it is during periods of rising prices and incomes that the greatest numbers of farmers have the money necessary to buy or make a down payment on a farm. Most farm purchases are partially financed with a mortgage, a real estate sales contract, or some similar form of credit instrument. If prices fall after a war, many farmers who contracted debts at wartime price levels find it difficult or even impossible to meet their obligations. The number of farmers who are in trouble and the seriousness of their situation largely depend on the extent of the price decline.

So far in World War II, as previously pointed out, the total debt of farmers of the United States has declined. It should not be overlooked, however, that while farm land values are still 30 per cent below the 1920 peak, and while large numbers of recent farm sales have been for cash or involve heavy down payments, it is still too early to be certain that the farm debt history of World War I will not be repeated.

AGRICULTURE AFTER WORLD WAR I

I should like to turn now to a consideration of price movements after the last World War and their effect on agriculture in the United States.

Prices of farm products continued up from the end of the war in 1918 until about the middle of 1920, when they dropped sharply. They reached a low about the middle of 1921, and then rose, stabilizing during the greater part of the following seven or eight years at a level approximately 40 per cent below the 1920 peak (figure 1).

What happened to agriculture during this period? As is usually the case in a period of falling prices, the prices of farm products dropped further and faster than farm wages and the prices of articles farmers buy, including interest and taxes. In the partial recovery after 1921, farm prices rose but did not get back into the prewar relationship with wages and other costs. As a matter of fact, the disparity between farm prices and the prices of articles farmers buy, as compared with the prewar relationship, continued on through the twenties and the thirties until World War II when farm prices again rose sharply.

It is obvious that a situation such as that which existed in the 1920's severely curtailed farm profits. Many farmers who were heavily in debt faced disaster. Many farmers who were not in debt were in difficulties. It is important to recognize, however, that all farmers were not affected alike. Whether we take the group who were in debt or the group who were free of debt, farmers producing certain products were harder hit than others. Farmers close to markets, as those in the Northeast, suffered a less severe price decline than farmers distant from markets, since transportation costs took out a smaller part of the price paid by the consumer. Furthermore, and perhaps most important of all, the fact is that farm land and farm units differ widely in their economic productivity and offer widely differing opportunities to reduce costs through the application of improved production techniques, including mechanization, and through increases in the size of the business.

Agriculture like other industries has made great technological progress in the past forty or fifty years. It has been estimated that the output per agricultural worker in the United States increased nearly three times between 1870 and 1940.[2] It has gone up materially since 1940.

Not all farmers have benefited equally, however, from the technological progress of the past fifty years. The development of hybrid seed corn, for example, does not benefit the farmer in the marginal corn lands of Nebraska and South Dakota, where he may plant hybrid seed but not get enough rain to produce more than half a crop in two or three years out of five, as much as it benefits farmers in the heart of the corn belt, where soil and climate are almost ideally adapted to corn production and where farmers can expect good crops year after year, assuming good management. Similarly, the farmer on a small farm unit or on poorly drained, droughty, or otherwise unproductive land finds it difficult and frequently impossible to make returns balance against costs when he invests in expensive modern farm machinery. I think it is a fair statement to say that

---

[2] Harold Barger and Hans H. Landsberg, *American Agriculture, 1899–1939: A Study of Output, Employment and Productivity* (New York: National Bureau of Economic Research, Inc., 1942).

the net result of technological improvements, including mechanization, has been to increase the disadvantage of farmers on unproductive land and farmers with small, uneconomic farm units. Farmers on good land and with farm businesses of larger than average size have had the advantage over the years, and their advantage has tended to increase. I should expect it to continue to do so.

I want to make it clear that when I speak of larger than average farm businesses, I am not referring to large corporation farms operated by gang labor, to plantations, or to groups of farms under a single management. These forms of ownership and operation are still relatively unimportant in the United States, and I think they are likely to continue so for reasons that I do not have time to go into here. When I speak of a larger than average farm business, I am referring to a larger than average farm operated primarily by family labor with perhaps some hired help—the kind of a farm business that workers in the field of farm management would refer to as a two- or three-man business. The actual acreage in such a farm will differ widely, depending on the type of farming. It might be as much as one thousand acres for a wheat farm, and less than twenty-five acres for a commercial poultry farm. Farm businesses of the kind which I have described are usually large enough to make efficient use of modern labor-saving machinery and equipment, and small enough to escape some of the inherent management difficulties of really large farms. It is on farms of this type that both the output per worker and the standard of living of all workers, family and hired, tends to be highest.

As I have pointed out, farmers on the less productive

land and farmers with small uneconomic units were at a distinct disadvantage in the 1920's in attempting to increase efficiency and reduce unit costs, a procedure which, of course, was one possible way to meet the unfavorable cost-price relationships which existed during this period. On the other hand, industrial activity was high and business good during most of the 1920's, and nonfarm jobs were to be had at good wages. The result was that this was a period in which there was a strong farm-to-city movement of population. The net migration from farms to cities during this period totaled nearly six million persons. Despite the fact that the farm population more than reproduces itself, there were nearly one and one-half million fewer persons on farms on January 1, 1930, than ten years earlier. As is usual in such circumstances, most of the migrants were young people. Sixty per cent of them were from the South where the rural birth rate is high and economic opportunities fewer than in the North, even in normal times.

Movements of farm population such as occurred in the twenties do not indicate that agriculture is decadent or that highly commercialized large-scale farming is taking over and that the family farm is on the way out. They represent a normal adjustment to a basic situation which is peculiar to agriculture. In progressive economies, such as that of the United States, the food and fiber needs of the population can be met by a decreasing proportion of the total population because of the relatively inelastic demand for these products and the fact that constant improvements in production techniques make it possible for one person in agriculture to turn out an increasing amount of product. In periods of rapid technological progress, like those we have had during the past twenty-five years, not only does the proportion of the total population required to produce food and fiber decrease, but there may be a reduction in the actual number of persons required.

A further characteristic of agriculture is that farm population more than reproduces itself. Since a declining proportion of the total population, and perhaps even a smaller number of persons on farms, is required to produce our food and fiber requirements, and since the farm population much more than reproduces itself, it follows that agriculture has an "exportable surplus" of population. There is a net migration from farms to cities in what we are pleased to call normal times. This flow is accelerated in periods such as the 1920's when the agricultural industry as a whole is in a relatively unfavorable position and when industrial and business activity, on the other hand, is relatively high, so that nonfarm jobs are readily available. The net effect of such movement is to shift surplus population off farms, take the less productive land and uneconomic units out of production, and increase the output per agricultural worker.

From the long-time standpoint of society, this is desirable providing we can keep the excess farm population usefully and profitably employed. It is by this process that the United States has raised its standard of living to the highest level of any country in the world. If agricultural efficiency in the United States were no higher than in most Asiatic countries, it would still require 75 or 80 per cent of our total population to produce our food and fiber requirements, leaving only 20 to 25 per cent to produce radios,

51

automobiles, refrigerators, and all the other things we like to think of as being associated with the American standard of living. In the United States at the present time less than 20 per cent of our total population is on farms. They are able to produce our food and fiber requirements with some available for export.

It is true that we have plenty of difficulties in our highly organized urban societies, including problems of periodic depressions with accompanying unemployment and distress. In such periods there is always a tendency to look to the land as a refuge from economic storms and to develop schemes for settling people on the land, usually under conditions that make their average productivity so low that they can never hope to enjoy what we are pleased to call an American standard of living. While we have problems in our modern industrial societies, we should not become confused and overlook the fact that in the last analysis our high standard of living is based upon high agricultural output per worker which makes it possible to release manpower to produce other things. Steps in the direction of settling people on unproductive land or on units too small to make efficient use of manpower and machinery represent a decrease in the total potential output of our economy and, therefore, in our potential level of living.

This is not to say that part-time farms and rural homes for city workers are undesirable. Personally I think they are. The point, however, is that they do not and cannot take the place of a regular pay check if the person living on a city lot in the country or engaged in part-time farming is to enjoy a high standard of living. In addition to a different way of life they represent a means of supplementing a

52

city pay check rather than a complete substitute for it.

Economic conditions such as we had in the 1920's had the effect, as I have pointed out, of bringing about certain shifts in population and adjustments in land use. In New York State alone about two and one-half million acres disappeared from farms, most of it land which because of soils, topography, climate, or location was relatively unproductive. Many small farms which were uneconomic units were combined into larger ones. The Census reported that the number of gainfully occupied persons engaged in agriculture in New York decreased 13 per cent between 1920 and 1930.

Even with nonfarm jobs relatively abundant, as in the twenties, and agriculture in a relatively unfavorable position, shifts in farm population tend to be slow and painful. When the individual farmer reaches forty or fifty years of age and is living on a farm for which he may be heavily in debt, it is difficult or even impossible for him to uproot himself and his family and move to another farm or shift to another occupation. He usually has had no experience other than farming and perhaps experience in only one type of farming. The result is that he frequently stays on his farm even though he and his family must accept a declining standard of living. When the children are grown they usually leave the farm if they can find city jobs or better opportunities in farming elsewhere. Only after the old folks pass on is the land abandoned for farming purposes.

Changes in land use on farms which continue to be operated are also difficult to make. Agriculture is a biological industry. The types of crops grown and the livestock kept are largely determined by conditions of soil, climate, and markets. From an economic standpoint, it is frequently difficult in specific situations to make adjustments to meet changing conditions. If a particular crop or livestock product is no longer profitable it may not be possible to find another product to replace it which will yield a satisfactory return. In such circumstances the old level of property values, debts, and taxes must come down if farmers and their families are to maintain past standards of living. Returns may be so low that the land eventually will be abandoned. Even if substitution of product is possible it takes time, frequently requires a substantial capital investment, and sometimes requires the learning of new skills. Livestock production programs have to be planned long in advance and crop rotations suitable for the maintenance of soil fertility cannot always be easily changed. The result is that even when adjustments are possible they are slow in taking place and farmers may remain in financial distress for years.

In periods like the 1920's, when farm prices are low relative to costs as compared with the previous period, large numbers of farmers are in difficulties. In a democracy when enough people get into difficulties there is usually agitation for the government to do something. This was true in the twenties. Numerous bills were introduced in the Congress, and some of them passed only to be vetoed. Others became law. Legislation providing for emergency feed and seed loans to distressed farmers was passed in 1921. The War Finance Corporation was set up in the early twenties to help bring about an orderly liquidation of debts incurred by the livestock industry during the war period. However, it was not until 1929 that the Federal Farm Board, which was one of the first important agricultural relief agencies, was established. We might describe the 1920's as a period in which economic conditions confronting agriculture were such as to cause deep rumblings, much dissatisfaction and agitation, but not serious enough to cause a sharp departure from the previously established policy of government of "letting nature take its course" in periods of agricultural depression.

Following the sharp break in prices which began in the early 1930's, conditions grew worse. Everyone who was in trouble before was in worse trouble now. Good farm businesses operated by good farmers were in the red, and even farmers with modest debts could not meet their obligations. Farm foreclosures increased rapidly. City jobs were not available, with the result that the normal farm-to-city movement of population not only slowed down but was actually reversed in many areas as sons and daughters who had lost their city jobs came back to live with the old folks.

In such circumstances, farmers demanded relief from the government and were able to get the support of other groups, who were also in difficulties, in getting relief. The question in the 1930's was not whether something would or would not be done, but what would be done. We compromised by trying a little bit of everything. I think it is fair to say, however, that if we had not had the New Deal and its attendant measures, we would have had some other kind of a Deal with some other kind of measures which I suspect some of us would have disliked just as heartily, since they too would have represented a greater degree of

55

governmental intervention than that to which we had been accustomed.

AGRICULTURE AFTER WORLD WAR II

I have discussed in some detail the history of economic developments in the field of agriculture during World War I and the subsequent postwar period for the purpose of pointing out the way agriculture reacts under different economic conditions. I suspect I may have gone down so many alleys and into so much detail that the whole thing seems thoroughly confused. I would like now to attempt to draw together the various threads and indicate some of the more important developments which are likely to take place in our postwar agricultural economy under different sets of assumptions.

First, let us assume that we have a further increase in the general level of prices after the war, that is to say, further inflation. Under such circumstances, further increases in farm prices would be expected and profit margins would in all probability continue favorable. The rate at which farm prices would rise and the margin of profits in farming would be largely determined by the speed of inflation. Land values would continue to rise, perhaps at an increasing rate, as it became increasingly clear that we were in for a real dose of inflation and as large numbers of investors began to invest in land as a hedge against inflation. If we should go on through to a grand crash, we would probably end up with large numbers of farms heavily encumbered with debts which would not be paid except on a compromise basis.

While we were on the way up, paper profits from farming would be high, and we would hear more complaints from farmers about governmental attempts to control farm prices, if such attempts were made, than we would hear about the need for some form of government assistance.

Let us make a second assumption. Let us assume that there is a further but moderate rise in the general level of prices during the period of reconversion after the war, but that we succeed in making the transition from war to peace without undue inflation and without excessive unemployment. Let us assume further that, after we get back into full production and goods are available in plentiful supply, the general level of prices declines somewhat from the wartime peak, much as it did after World War I, but prices are then reasonably stable for a period of several years at a level below the peak but above prewar levels. If such a situation should develop, I hope that we could avoid the sharp drop in prices and the accompanying period of declining business activity, unemployment, and sharply falling returns to agriculture that occurred in 1921. In other words, I am postulating a situation essentially similar to that following World War I with the sharp dip in prices between 1920 and 1922 at least partially eliminated.[3] What would be the effect on agriculture of such a situation?

Despite shortages of farm manpower and materials, agricultural production during the war years has been at high levels. We have not only provided a diet for our civil-

---

[3] From a purely economic standpoint the surrender of Japan, which occurred after the above was written, increases the difficulties of making a smooth transition from war to peace. It appears now as though we would have considerable unemployment during the reconversion period.

ian population which on the average is equal to and perhaps in some respects better than the prewar one, but we have provided the food necessary for a large fighting force and have exported substantial quantities under Lend-Lease. It is true that a considerable part of our phenomenal wartime farm production has been due to a series of favorable growing seasons. It has been estimated that weather accounts for from one-quarter to one-half of our increased wartime farm production. Even after allowance is made for unusually good weather, however, and even if we assume full employment and high consumer incomes in the postwar period, it looks as though there would be enough pressure of farm production on prices to force farm prices down relative to wages and to the prices of nonfarm goods, which will be in short supply. This assumes that the export market will absorb only a limited quantity of United States farm products at prices that are satisfactory to farmers after the initial period of getting food production reestablished in war-torn countries is passed. This period is likely to be shorter than we realize. At least this was the experience after World War I. While diets in many European countries and in Asia will continue to be inadequate for large segments of the population, I doubt if we shall continue to export any substantial quantities of food to these groups, since they are not likely to have the necessary purchasing power to pay for it. A need will exist in terms of nutrition, but nutritional need is one thing and effective economic demand backed up by purchasing power is another.

The extent of the decline in farm prices relative to the prices of nonfarm products will obviously be affected by

58

conditions of demand in this country and by the extent to which export markets actually open up after the war. If we have reasonably full employment and a high level of consumer incomes, the gap between farm prices and nonfarm prices and wages will be less than if we have widespread unemployment as we did in the 1930's. Similarly, if there is a substantial volume of international trade resulting in a high level of business activity in other countries, we can expect to export more food and fiber at satisfactory prices than would otherwise be the case.

If farm prices decline relative to nonfarm prices and farm wages in the postwar period, we shall have price-cost relationships in agriculture similar in character to those which existed during most of the period 1922–1930. If this should occur, it is reasonable to assume that much the same problems will develop. For one thing there will be heavy pressure on farmers to reduce costs. Some farmers will be able to do so by increasing the size of their businesses to gain greater operating efficiency, by increasing the use of machinery, and by increasing output per acre and per animal. For various reasons, other farmers will not be able to make the necessary adjustments with the result that their incomes will be sharply reduced. Farmers who purchased land during the war at high prices and entered the postwar period heavily in debt will be in trouble. So will farmers on land which is relatively unproductive and on farm units of uneconomic size. Under similar conditions in the 1920's, we had a large migration of farm people to cities, towns, and villages.

We have already had a heavy farm-to-city migration during World War II. According to estimates of the Department of Agriculture, 25,190,000 persons were living on the farms of the United States in January, 1945. This is the smallest number of farm residents in the entire thirty-five-year period for which estimates of the farm population are available, and represents a decline of approximately five million since January, 1940. This is a decrease of nearly 17 per cent. The migration of farm people to cities, towns, and villages as a result of the wartime expansion in non-agricultural employment has been the chief factor in the decrease of farm population since 1940, although enlistments and inductions of young people living on farms has also been an important factor.

If we maintain a high level of industrial activity after the war with plenty of jobs, and if agriculture is in a comparatively unfavorable situation as it was after World War I, it is probable that only a relatively small percentage of the persons who left farms during World War II will return to them after the war. As a matter of fact, we might have further acreages of our poorer land going out of use for farming purposes. Such land may not be abandoned, however. Many farm people who have taken city jobs during the past ten years have not moved off their farms, as they did in the 1920's, if the farms were located within reasonable driving distance of nonfarm employment. Improved highways and automobile transportation make it possible to live a considerable distance from one's job. This was not the case during most of the 1920's, when it was necessary to change residence in order to change jobs. In most parts of the Northeast and in many other parts of the country it is now possible to shift from farming to a nonfarm job without changing residence. It seems probable that in the postwar period we will continue to have substantial increases in our rural nonfarm population. Not only will a certain percentage of farm people transfer to nonfarm jobs if employment opportunities are plentiful, but city people probably will move to the country in increasing numbers. This prospective intermixture of farm and nonfarm groups has far-reaching social and economic implications which are beyond the scope of this discussion.

What about government farm programs after the war? In the 1920's we had a great deal of discussion about farm relief, but relatively little was done. However, since our experience in the 1930's we shall undoubtedly have various programs to aid agriculture, even though general economic conditions may prove to be relatively favorable. Once started, programs of this type tend to persist and to expand rather than to contract. As a matter of fact, we already have legislation on the books providing for support prices for farm products at 90 per cent of parity and at 92.5 per cent of parity in the case of cotton. This price-support program is to continue for as long as three years beyond the cessation of hostilities.

Now for the third and last assumption. What if we should have severe deflation after the war, with prices declining to or below prewar levels? With all of the excess purchasing power now in the country and everyone concerned with the problem of maintaining full employment after the war, such a course of events does not seem probable. If it should occur, however, the situation that existed in the early 1930's would probably repeat itself so far as agriculture is concerned. I suspect that under such circumstances, if the depression lasted long enough, we would

61

end up with an almost completely regimented economy. In the case of agriculture, there would be widespread demands for price supports of all sorts. Although they might ease the shock, they would not prevent declining prices. We could, however, look for minutely regulated programs of production control since production control goes with government support prices just as rationing goes with price ceilings when food or other commodities are in short supply.

Economic developments in the economy as a whole will obviously be the most important factor determining what happens in agriculture during the postwar period. However, we do not know what these developments will be. For this reason I have attempted in the foregoing discussion to point out the way agriculture reacts to changing economic conditions rather than tried to forecast what economic conditions will be after World War II. Economic developments in the postwar period will be dependent upon a large number of factors including monetary, credit, and fiscal policies; government policies in other fields; political and economic developments in other countries; and the policies of business, labor, agriculture, and other groups. All one can hope to do in trying to forecast future economic developments is to attempt to analyze the various forces now at work or likely to enter the picture and to appraise the probable net effect of these forces on our economy. Such an appraisal is always subject to wide errors of personal judgment.

I would like to make a closing comment which it seems to me applies to all major groups in our economy—business, industry, labor, and agriculture. In a competitive economy it is to be expected that each individual and group will be out to get as large a slice as possible of the cake represented by total national product. It seems to me, however, that in recent years almost every group has emphasized the use of restrictive measures to try to improve its economic position. In other words, each group has attempted to get a larger piece of the national cake, even though the methods used had the effect of substantially reducing the size of the cake as a whole. Agriculture restricted production in the 1930's when millions of people were hungry, only to find that a high price per unit of product times a small output, brought about by production control programs, gives approximately the same answer as a lower price per unit of product times a larger output.

No reasonable person objects to good wages and fair hours for labor. But when output is restricted to substantially less than a man can produce in a day, without injury to health or excessive fatigue, it seems to me that the rest of society is at least entitled to raise a questioning eyebrow. The same goes for industry, which sometimes restricts output to maintain prices, when, like agriculture, it might find that a lower price per unit of product times a larger output would give substantially the same answer in terms of earnings as a higher price times a limited output.

The only way we as a nation can enjoy a higher average level of living is by increasing the total national output, a point we seem to have largely overlooked in the 1930's. It is important that we keep this elemental proposition clearly in mind in the years ahead when there undoubtedly will be moves to freeze efficiency at existing levels through restrictive programs, some of which will undoubtedly be instituted by the government in the interest of helping this or that disadvantaged group. Measures to assist disadvantaged groups are in order if it is clear they cannot keep themselves. It is important, however, that the measures used be such as to bring about desirable long-time adjustments rather than to permanently freeze situations which involve the inefficient use of manpower, land or other production resources, and which therefore constitute a drag on the economy as a whole.

THE ROLE OF ORGANIZED
LABOR IN THE
UNITED STATES ECONOMY

*

BY LOUIS HOLLANDER

PRESIDENT OF THE NEW YORK STATE CIO COUNCIL

THE TREACHEROUS Japanese attack at Pearl Harbor on December 7, 1941, confronted the American labor movement, like every group in our nation, with its greatest test. There was an immediate response from organized labor. Scarcely had the impact of this tragic event reached the far corners of the United States when labor's unconditional pledge of support was on President Roosevelt's desk in the White House.

In messages, Philip Murray, President, Congress of Industrial Organizations, and William Green, President, American Federation of Labor, pledged that millions of organized workers would do their utmost in the fight against the outrageous aggression of Japanese imperialism and would help secure the final defeat of the forces of Hitlerism.

With full realization of what total war means, labor turned grimly to the task of winning the war. Organized labor was rededicated to the cause of democracy. Organized labor enlisted in this cause and has fought for its full realization almost from the beginning of our history as a nation.

The emergence of the trade union movement in the United States is in itself an American epic. The first fumbling effort took place in 1792, only three years after the adoption of our Constitution. It was then that the first American trade union was organized in Philadelphia.

The course of development has not been smooth or easy. In those early days, the mere organization of a trade union was banned as a conspiracy. Men and women were sent to jail for the crime of "restraining trade" by attempting to organize their fellow workers for the purpose of bettering their working conditions.

Every advance recorded since those first beginnings has been achieved as the result of struggle. The history of organized labor in the United States is one of constant struggle against such devices of employers as the lockout, the yellow-dog contract, the blacklist and labor injunction, private police, strikebreakers, and even the use of the militia and the United States Army to break the rising trade union movement. The history of our country's social and economic development is closely interwoven with the story of labor's great strike struggles: the Haymarket riot in Chicago, the Carnegie Steel strike in Homestead, Pennsylvania, the Pullman strike, the Lawrence and Patterson strikes, and the great steel strike of 1919. Nor should we forget more recent events such as the Little Steel strike of 1937 with its shameful record of ten workers shot and killed on the South Chicago prairies on Memorial Day.

The struggle for the right of self-organization has always been closely interlinked with the struggle for social legislation to guarantee basic minimum rights to the workingmen and women of America. Our progress in this field during the last one hundred years can best be indicated by the fact that the first state law relating to hours of work, enacted in Massachusetts in 1842, set a *maximum ten-hour day for children under twelve,* leaving the hours of labor for all others—men, women, and children—*unrestricted and unlimited.*

Those of us who have spent a lifetime in the labor movement can look back, in our own experience, to the time when the seven-day week and the fourteen-hour day were normal working conditions. We can remember, too, the days of the four-dollar-a-week wage, when the worker was expected to bring his own machine to the shop and to pay for the gas which he used in doing his job.

Yet, with all these years of effort and struggle behind it, the American labor movement, in 1933, still found itself numerically small and relatively weak. It was under the first two administrations of President Roosevelt that for the first time the economic and social rights of the workingmen and women of America were given full recognition and written into law. Under these administrations, the National Labor Relations Act and the Social Security system were placed upon our statute books and now have become accepted as an integral part of our American way of life.

It has been in the last twelve years, too, that the organized labor movement has experienced an unprecedented growth. Membership has increased by almost fivefold from little more than three million in 1933 to nearly fifteen million in 1945. Even more significant is the fact that the great mass production industries—steel, automobile, and aircraft; rubber, electrical, textile, and machine; and shipbuilding—remained wholly unorganized as late as 1935. It was only with the advent of the CIO in that year that a serious effort was made to organize these key mass production industries. These were unionized under the leadership of the CIO in the course of militant struggles marked by large-scale, bitterly fought strike actions between 1935 and 1940. These new unions had scarcely succeeded in establishing themselves on a secure foundation when the war crisis confronted labor and the nation with tasks which overshadowed all others.

It is to the everlasting credit of our labor organizations that they reacted to this crisis with none of the immaturity of youth. When the great test came, they proved themselves as capable of statesmanship and restraint as they had of militancy and determined struggle.

Placing national unity for victory above any narrow or selfish interests, labor has sought no special advantage for itself. It has at all times subordinated its own demands to the greater needs of a nation at war. At the same time, it has vigorously pressed for the adoption of those economic and social measures required for maximum production. It has had to learn how to fight for the adoption of these measures in a manner which would advance, not retard, the war effort.

Immediately after Pearl Harbor, organized labor voluntarily pledged that it would not strike for the duration. So loyally has that pledge been maintained that less than one-tenth of one per cent of working time has been lost because of strikes. Instead, labor has found and applied new techniques, substituting the processes of conferences, public pressure, and political action for the strike and the picket line.

It also had to conduct a struggle against those "business-as-usual" industrialists who were more concerned with their profits than with the defeat of the enemy, and who tried to provoke labor into hasty and ill-considered action, hoping thus to discredit, weaken, and ultimately destroy its organizations.

In the initial stages of the war production program, organized labor played a leading role in formulating and pressing for the execution of programs to make the conversion from peace to war production speedy and complete. It has been in the forefront in retraining millions of workers as well as in bringing people into industry who were wholly without factory experience. Labor has been particularly active in insisting on the full utilization of women and Negroes and in fighting against all forms of discrimination against them, whether in the form of unequal wages or in the denial of access to higher skills.

The trade unions have also concentrated on increasing labor productivity. Through labor-management committees and in other ways, workers have come forward with thousands of proposals to increase productivity. Today, with approximately one-fifth more workers employed in industry than in 1939, American workers are turning out more than twice the volume of production of that year. A part of that increase is accounted for by the longer working day, but a large part is the result of more intensive effort and improved technique, for both of which labor is entitled to the major share of the credit.

The role of organized labor in our nation's war effort is written in the record of its magnificent contribution to the miracle of American production. Shortly after Pearl Harbor, President Roosevelt set our war production goals. These were sneered at by the Axis and ridiculed by our native doubters. In a challenge to America, President Roosevelt called for sixty thousand planes, forty-five thousand tanks, and eight million tons of shipping.

What does the record show? What made possible the tremendous pattern bombings which destroyed German industry? What made possible the tremendous onslaught on the coast of Normandy? What made possible the relentless advance of the Russian armies after Stalingrad? What made possible the maintenance of the steady surge of our Pacific forces against Japanese imperialism?

All this was made possible by the tireless and devoted effort of American workers who produced not the sixty thousand planes called for by President Roosevelt, but over two hundred and forty-six thousand planes. To meet the challenge of eight million tons of shipping American labor produced forty-five million tons and at the same time constructed over fifty-six thousand naval vessels of various types.

Yes, it is true that statistics are sometimes tiresome. But it was the two and one-half million machine guns, the six million rifles, the over five million carbines, the two million submachine guns, the seventy-five thousand tanks, the one hundred and thirty thousand self-propelled guns, the fifty-six thousand pieces of field artillery, and the hundreds of thousands of trucks and jeeps which gave to the armies of democracy the mightiest aggregation of fire power in the military history of the world.

The record of American workers on the production line is only excelled by their service on the battle line. Approximately three and one-half million members of trade unions have served and are still serving in the far-flung corners of the world, where they have contributed their share of the gallant and heroic exploits of this war.

But the labor movement was not only active in achieving what has been justly termed "the miracle of production" and in sending its sons and daughters into the armed forces. Labor has played an important role in every field of effort connected with the war. Labor has been a generous contributor to every war fund. The men and women of labor have been among the most frequent patrons of the Red Cross blood bank, and labor has done its share and perhaps even more in the various war loan drives.

Since December 7, 1941, between the Congress of Industrial Organizations and the American Federation of Labor, more than 205 million dollars was raised for the various war funds, the Red Cross, and local community chests. In 1944 alone the two organizations raised sixty million dollars for the National War Fund.

As for the war loans, the figures are almost unbelievable. It is conservatively estimated that labor purchased five hundred million dollars worth of bonds in the recent Seventh War Loan drive. Perhaps the best idea of labor's contribution to financing the war may be gained from the following figures. In December, 1941, only seven hundred thousand workers throughout the nation were purchasing war bonds on the payroll savings plan. Today, approximately twenty-six million workers are enrolled in the payroll savings plan. They buy five hundred million dollars' worth of bonds each month as well as extra bonds during drives.

But this total war is not being won only on the battlefield and on the production line. Victory over the enemy and victory for the things for which we fight also depend on the correct solution of fundamental political questions. So labor has also had to turn its attention to politics.

Labor recognizes that there are powerful and unscrupulous forces at work in our own country that fear a free and a united people. Their program calls for a regime of reaction at home and an era of aggressive imperialism abroad. They do not want peaceful co-operation between capital and labor in the United States. Instead, they seek to deprive labor of the gains which it has won and smash its organizations. They do not want international co-operation abroad. Instead, they seek world domination and dream of an "American Century."

These forces scored significant gains in the Congressional elections of 1942. Encouraged by this success, they redoubled their efforts to undermine public confidence in President Roosevelt. They obstructed legislation necessary to strengthen and solidify the home front, and sharpened their attacks upon labor and the progressive forces. In this way, they prepared the ground for an all-out bid to take power in the national elections of 1944.

In July, 1943, the leadership of the CIO, taking note of the developing crisis, organized to meet it. On the initiative of President Philip Murray, the CIO organized its Political Action Committee, under the chairmanship of Sidney Hillman. The Committee at once set about the task of organizing labor and progressive forces for independent action in support of the win-the-war and win-the-peace policies of President Roosevelt. By every means at its command, it brought the decisive issues of the election to the

74

people. It presented a program for national unity to win the war, for international co-operation to assure an enduring peace, and for a domestic economy to yield full production and full employment in the postwar world.

Knowing that a large vote would be a progressive vote, it devoted its skill and energy in organizing to the task of getting the largest possible number of voters to the polls. The Committee taught the simple truth that, under our democratic form of government, political power rests with the people. Exposing the campaign of demagogy and confusion waged by the camp of reaction, it showed the people how to exercise their power fully and intelligently.

As an unwitting tribute to the influence of the CIO and its Political Action Committee, the reactionaries concentrated their venom upon it. By every known device, they attempted to isolate it from the main body of the American people, to set the people against it, and so to destroy it. Had they succeeded in their desperate effort, conclusive and uncompromising victory over fascism would have been endangered. And we should have lost—in our generation at least—all hope of establishing a lasting peace.

They failed. They failed because the Political Action Committee voiced the hopes and aspirations, not of labor alone, but of all progressive Americans, and because it gave its full co-operation and support to all those who were traveling along the same road. Thus it took a leading role in building, and became an integral part of, a broad people's movement, embracing the rank and file and the progressive leaders of the American Federation of Labor, farm groups, church, business, and professional groups, and the progressive elements in both major political parties.

And because they failed, the United Nations Charter is today an accomplished fact. Labor played a major part in mobilizing the sentiment of the American people in favor of a world organization that would insure a permanent peace in the world.

Far from describing this ideal as impossible of attainment, labor pitched in to make this dream a reality. The sentiment of America's workers, their bitter opposition to war and their desire for lasting peace, was crystallized by the CIO Political Action Committee. Meetings were held all over the nation, and the voice of the American people was heard in San Francisco.

Today, with the Charter an accomplished fact, signed by some fifty nations of the world, there are not only vague hopes, but the first concrete steps have been taken in the direction of permanent peace in the world.

Now V-E Day has come and gone and we are bending every effort to hasten the approach of V-J Day and the final defeat of the enemy. Yet even as we concentrate all our energies to speed the day of victory over Japan, the problems of peace make urgent demands upon our attention. As reconversion to peacetime production begins, another supreme test of our democracy is at hand. We were able to unite all of our efforts to achieve military victory. Will we be able to remain united for total peace as we have for total war?

On behalf of labor, I can say that we are willing and eager to remain loyal and energetic partners of all other progressive Americans to realize the great goals which we share: a peaceful and prosperous nation in a peaceful and prosperous world.

These great goals can be achieved only if we realize full employment at home, the freest exchange of goods and services with other nations, and a continued rise in the living standards of all the peoples of the world.

Here at home we have achieved full employment for war. We have doubled our national income and provided a job for every man and woman able and willing to work. Thus the old argument that full employment is an illusion has been exploded. The great challenge which faces us today is whether we can convert our capacity for the manufacture of instruments of war to the satisfaction of the needs of a world at peace.

Let us not minimize the difficulties involved. V-J Day will return some ten million men and women from the armed services to civilian life. It will release another fifteen million from war production. Some twenty billion dollars in new plants and equipment will have been added to our industrial capacity. Yet the major contribution of government to our national income, in the form of war contracts, will largely cease.

On the other hand, the need of the men and women of America and of the people of the whole world for the goods and services which we are geared to produce can provide our industries with work and our people with jobs.

Americans need more food. In 1942, Paul V. McNutt, then Federal Security Administrator, estimated that forty million Americans lived on a diet dangerous to their health. The production of sufficient food to give every citizen an adequate diet will guarantee our farmers a decent income and provide work for tens of thousands of men and women in the production of farm machinery and in the food and food-processing industries.

Americans need better housing. Almost one-half of all dwellings in the United States are in need of major repairs or are without baths or electricity. At the present time, in our cities alone, more than thirty million people are living in homes that structurally or because of their location in slums are below essential minimum standards of health, safety, and decency. In the farm districts of the South only 9 per cent of all farm houses have inside running water. It is commonly acknowledged that more than one-half million nonfarm dwelling units per year are required if only to provide for the natural increase in the number of families and in order to maintain present standards. Experts tell us that we must build one and one-half million new farm and nonfarm units each year for the next fifteen years in order to put our housing situation in order. This will provide a source of tremendous employment opportunities for many years to come.

We need more and better schools. It is a sad fact that there are over ten million illiterates in our country. More than one-half of all our public school buildings contain but a single room. Today, when our democracy is undergoing a major attack from the forces of nazism, fascism, and Japanese imperialism, full education for our people is vital to the continued health of our free institutions. We must provide education for all our people—regardless of race, creed, color, or national origin. And we must do away with the disparity between rural and urban education. To meet these needs, and to attain the goal of providing every adult American with at least a high school education, will provide countless employment opportunities in every field of education.

The magnitude of the job which confronts us in this field may be realized from the fact that as late as 1940 some three-fourth million children of elementary school age and more than two million of high school age were not in school at all, primarily because of low family incomes and the necessity to add to those incomes. Those enrolled in educational institutions above high school level represent only a small fraction of the number who have the mental capacity to continue their education in college. Again this is primarily an economic question.

In order to revamp our system of elementary and secondary public education completely and in order to include allowance for attendance at kindergartens, nursery schools, and junior colleges, the research division of the National Education Association has estimated that we would have to provide for an increase of more than five million in the student enrollment and of nearly 50 per cent or about four hundred thousand in the teaching staff.

American health is bad. One-third of the men examined for the Army were rejected because they could not meet its physical requirements. To assure a healthy nation, we will need three times as many doctors and dentists as we had before the war. We will need hundreds of new hospitals and hundreds of thousands of nurses, research and laboratory workers, and other personnel in the medical and dental fields. However, labor realizes that neither labor nor industry alone can solve the tremendous health problem of the nation. All the energies of the government must be called upon to meet the health problem of the American people. A number of labor organizations have established health protection for their members through collective bargaining agreements, but this includes only a relatively small group and therefore does not affect the American people as a whole. Labor is therefore wholeheartedly supporting the Murray-Wagner-Dingell health bill, which will create a comprehensive social security system giving larger benefits and offering protection to some fifteen million additional workers who are not covered by the present social security act. As introduced in the Senate, the bill increases the amount of unemployment compensation benefits and creates a national social security system including old age retirement and survivors' insurance, disability insurance, lump sum death benefits, unemployment and temporary disability insurance, maternity benefits, and prepaid personal health service insurance.

Under this bill, workers suffering from unemployment and those temporarily disabled will receive up to thirty dollars a week; married women workers will receive twelve weeks of maternity leave. All workers covered by the act will be entitled to the services of a physician and will be able to chose any physician in the community who is part of the system.

We need more autos, more and better roads, more railroads, planes, and every type of transportation and communication. Our people need more clothing, more iceboxes and radios, more of everything that goes to make up a good and abundant life.

These needs and the needs of the people of other nations

provide us with a new economic frontier as challenging as the geographic frontier which confronted the American pioneers. This new frontier cannot be conquered by wishful thinking or by the haphazard efforts of individuals. Bold planning and the unselfish co-operation of all groups is essential to its achievement. Labor has such a plan.

The CIO Re-Employment Plan, for example, is mainly a program for industrial production and employment. Let me recite some of its features.

The CIO plan is concerned first of all with the buying power of the workers because the lack of that buying power is the main source of our economic ills.

This is the field in which organized labor has a peculiar responsibility and a special obligation to speak out. And not only for union members. Throughout the program the CIO calls insistently for the protection and welfare of all the people.

Labor is deeply interested in the security and prosperity of every section of the nation and wants real prosperity for the farmers. Labor is concerned with the problems of independent businessmen and professional people and is vitally interested in the welfare of returning veterans. The CIO champions the cause of all racial and national minorities.

Every veteran of both this and the earlier war must have his job opportunity in a national program of full production and full employment.

The same is true of the enlarged group of women who from necessity or choice will be in the labor market when the war is won. Women must not only have democratic employment opportunities; they must receive equal pay for equal work.

The Negro worker has given his efforts to production for victory; his employment also must be without discrimination in an expanding economy to which all can contribute their best efforts and from which all can obtain an adequate living.

The prosperity of the industrial worker is found only in the prosperity of all Americans. Each of us is secure only as all of us are secure.

The CIO has outlined its views and policies on all these questions in programs which it has issued from time to time. It offers its friendship and co-operation to every group among the American people.

The solution for most of our common problems, however, is full industrial production and employment. Farmers, independent businessmen and professional people prosper as the total take-home pay of industrial workers rises. There need be no conflict between races and national groups, or between veterans and workers, if we make sure that there are jobs for all. CIO offers the following six-point program:

(a) There must be immediate revision of the Little Steel formula to compensate labor for the loss incurred owing to the rise in the cost of living. In due time labor must, in addition, share more fully in the earnings of industry through further wage increases.

(b) There must be prompt and generous increases in wage rates to protect workers against reconversion unemployment.

(c) There must be no reduction in take-home pay as overtime is eliminated in war plants.

(d) Annual wage guarantees must be included hereafter in labor contracts.

(e) Dismissal pay, sick-leave pay, paid vacations and holidays, paid insurance, veterans' funds, and the elimination of geographical differentials must be elements of collective bargaining contracts in the future.

(f) There must be an unqualified acceptance of the principles of collective bargaining.

Nor can we realize our goal of jobs for all if the purchasing power of the American worker is permitted to sink back to prewar levels. In that direction lies the downward spiral of idle men, unemployment and want, and a depression worse than the black days of 1929. Today, when our national income has reached its highest peak, one out of every three American families still earns less than one thousand dollars a year.

The Congress of Industrial Organizations has made the elimination of substandard wages one of its main concerns. At its 1944 convention a resolution was passed calling for the enactment of legislation to declare all wages of less than sixty-five cents an hour substandard. Labor is, therefore, supporting an amendment to the Fair Labor Standards Act to change the minimum hourly wage rate from forty cents to sixty-five cents, to go into effect immediately, and to raise it to seventy-five cents an hour in 1947. While the CIO is supporting the Pepper resolution of sixty-five cents an hour as a temporary war measure, the real aim of labor is to raise the minimum wage under the Wage and

Hour Law from forty cents an hour to seventy-five cents an hour.

There is much talk by business-minded orators about the "American way of life" and the "American standard of living." It comes as a bit of a shock to find that in 1941 over one-half of all American family units had a total income of less than fifteen hundred dollars a year; that one out of every three families had less than one thousand dollars a year income; and that one out of every six had less than five hundred dollars a year income. The war boom has somewhat improved the picture. An OPA study made in 1943 showed that "only" one-third of our families now have incomes of less than fifteen hundred dollars a year. Not only is this a temporary wartime phenomenon, however, which will vanish with the coming of peace, but the rise of more than 40 per cent in the cost of living during the war makes a lot of this gain a statistical illusion. If one looks merely at average income figures, it appears that the orators may be right. In 1943 the average income of a single consumer was $1,884 and that of a family was $3,481. This looks very pretty until we examine the top and the bottom layers of the social scale and we discover that at the top there is too much money perhaps for the number of people while at the bottom there are too many people for the amount of money. Thus, the figures show 15 per cent of the families with incomes of five thousand dollars or over that year received 43 per cent of the total income, whereas 30 per cent with incomes less than fifteen hundred dollars a year split up between them only about 8 per cent of the total income. The CIO is therefore directly interested that substandard wages be eliminated in all fields whether unionized or not unionized at present. No labor organization can take a narrow view of this great problem, which will only be licked through the united efforts of all workers from the strongly organized industrial workers to the underpaid white-collar clerk.

Let us see who are the "substandard." I quote here from the pamphlet issued by the Research Department of the National CIO. The wage earners whose toil yields wretched rewards are scattered throughout our economy. Some of them are in domestic service; a great many are farm workers; some are government employees. Since these three groups do not come under the jurisdiction of the War Labor Board, even the sixty-five-cent minimum resolution introduced by Senator Pepper will not apply to them. Of the 27,710,000 wage earners in industry, mining, transportation, trade, and services who are covered by the resolution, no less than 10,171,000 or 37 per cent are now earning less than sixty-five cents an hour. Those who boast about the "American standard of living" as an aim already achieved should be reminded of the fact that in the peak year of the present war boom, when the national income is at an all-time high, almost one-half of the workers in American trade and industry are still not earning enough to meet living costs.

I am confident that we in America possess the energy, the resources, the initiative, and the creative imagination to solve the problem of full production for peace, just as we proved ourselves capable of the miracle we wrought in producing for war. Our amazing record of war production was the result of the combined effort of all groups—workers, farmers, and businessmen—mobilized and united by the crisis which faced our nation. By common consent it was government—our government—which took the initiative in organizing and co-ordinating our common effort. So, too, I believe that the problem of full employment in peacetime can be solved by the same kind of joint effort. Industry, labor, and agriculture, working and planning together in partnership with, and with the full assistance of, government, can do the job.

In facing this task, labor is deeply conscious of its obligation and that of the whole nation to the men and women in the armed services who will soon return to civilian life. They have made tremendous sacrifices for us all. They properly expect that the country for which they fought will provide them with jobs and security upon their return.

These are not strangers to labor who are coming home. More than three million come directly from our ranks, and other millions are the sons and daughters of our members. Most unions have already made contract provision for the return of their members in the armed forces to their peacetime jobs, with accumulated seniority during the time of their military service. Many labor organizations have waived initiation fees for veterans who desire to enter an industry. Labor has been in the forefront of the fight for legislation for veterans' benefits, with free educational opportunities, retraining, and care for the returning servicemen.

But labor recognizes that provisions like these are not enough. Only if we succeed in realizing our goal of full employment, with a job for every American able and willing to work, will we have discharged our obligation to the men and women who fought and died for our freedom.

Another evil that must be rooted out from our midst, if we are to succeed in our plan for a more perfect political and industrial democracy, is the evil of discrimination which is based on race, creed, color, or national origin. As long as we have discrimination there can be no such thing as full employment or a genuine prosperity. The weapon of discrimination is one that the enemies of democracy have used to divide us. While one group among us is deprived of the benefits that accrue to others, there can be no genuine democracy. Our Federal Constitution forbids discrimination, and yet we know it exists. And this is true of every section of our country, not of one section alone.

When our need for manpower became acute during the war emergency, a Fair Employment Practices Commission was created by our Federal Government. This must not be allowed to die with the end of the war. We must carry out the provisions of our Constitution and end discrimination in employment, in housing, in education, and in every other field where it exists. No one should be denied the right to a job for which he is qualified merely because of creed, color, race, or national origin. We cannot legislate against prejudices. Unfortunately they do exist. We can, however, and should, legislate against discrimination that deprives a considerable section of our population of the right to a job. As long as millions of our fellow Americans are kept in poverty and ignorance, there can be no true democracy, no true prosperity, for the rest of us.

In our own state, we have taken a long step forward with the enactment of the Ives-Quinn antidiscrimination in employment bill and with the appointment of the commission to carry out this measure. This was no easy accomplishment. The forces of reaction in the state united to oppose the bill. The New York State Chamber of Commerce had the gall to threaten us with possible race riots, pogroms, and other evils if the bill was enacted. This sinister attempt at blackmailing the legislature into throwing out the Ives-Quinn bill was repudiated by labor, religious, and progressive organizations as well as enlightened business groups in our state. The Chamber of Commerce, as reported by the *New York Times* of February 12, 1945, in a letter to Governor Dewey and to all the members of the legislature, stated that this bill if enacted will make New York State "an undesirable place for employers and will bring burning resentment which would result from enforced employment of undesirable persons and would furnish fuel for intolerance and tend to foment rather than to eliminate the possibility of race riots, pogroms and the evils associated with the Ku Klux Klan and Silver Shirt organizations."

This was the most brazen statement ever made by any organization. Perhaps this statement by the Chamber of Commerce rallied all the liberal and progressive forces behind the bill. The CIO took the lead and mobilized all labor organizations in this state for the bill. In fact, to quote the *New York Times*, "The CIO has virtually made this bill its own." We led in the fight, not on a narrow partisan basis, but in the interest of all the people in the state and in the nation.

On the national scene the Fair Employment Practices Commission was struggling for its very life. We had the sad spectacle recently, when it came up for renewal in Congress, of the Rankins, the Bilbos, and others of their kind raving in their mad desire to intensify instead of end racial conflict. These men have forgotten nothing and have learned nothing. In their speeches they voice Goebbels' philosophy of racial supremacy. They attack everyone who does not agree with their own narrow, prejudiced views. They do not hesitate to pour out their poison on racial minorities and even on religious leaders who support the FEPC bill.

In spite of the poll taxers, a successful compromise has been reached on the FEPC in which the agency gets two hundred and fifty thousand dollars without a "dead sentence" clause. This was the result of hard work, good planning, and able leadership. The hard work was done in Washington and in the field where thousands of people poured wires and letters upon the Congressmen, demanding continuation of the FEPC. Public pressure was so strong that even the poll taxers of the Senate like Bilbo and Eastland were finally shut up—though not until they had done plenty of damage to national morale by their filibustering.

The same public pressure that forced this week's acceptable compromise on the FEPC must be applied too to the poll tax repeal bill, now and throughout the summer recess. Repeal of the poll tax will be the answer to the Bilbos, the Eastlands, and the Rankins. It will prevent such disgraceful performances as that of the last two or three weeks from being repeated in the future. So, the Fair Employment Practices Commission that was created by Franklin D. Roosevelt was not permitted to die.

Now labor is facing another great struggle. It is almost ten years since the Wagner Act became a law. The Wagner

Act, which has well been called the Magna Charta of American labor, established by law the basic right of American workers to bargain collectively, to be represented by unions of their own choosing, and to hold impartial elections at which the workers may choose their own representatives, and compelled the employers to negotiate with these legally elected representatives. The Wagner Act also protects the workers' rights from interference by their employers.

It is no exaggeration to say that the Wagner Act has been one of the most important factors in stabilizing our industrial and labor relations. In ten years the National Labor Relations Board has handled more than seventy-four thousand cases involving the right to collective bargaining. More than eleven thousand formal decisions have been issued and sixty-two thousand cases handled without formal decisions or court action. More than two thousand company unions have been disbanded since 1935, and three hundred thousand discharged workers have been reinstated. More than thirty thousand workers received some nine million dollars in back pay. Twenty-four thousand elections were held during this period, 72 per cent of them by agreement of both sides, and six million workers voted in these elections to choose their representatives.

Having failed in their attacks on the Wagner Act, the employers are now trying to kill the Wagner Act by means of the so-called "Federal industrial relations bill" which was introduced by Senators Burton of Ohio, Hatch of New Mexico, and Ball of Minnesota. Reactionary employers never accepted the basic principles of the Wagner Act, and they have never given up their constant battle to amend or even nullify the National Labor Relations Act. The Wagner Act has been under fire continuously since 1935, when employer associations advised their members not to obey it. When the Supreme Court in 1937 upheld the Act, the employers shifted their attack to the halls of Congress and tried in every way to undermine it.

No matter what the sponsors claim, the Ball-Burton-Hatch bill is actually an attempt to destroy the American trade unions and to enslave labor. This bill sharply restricts the right to strike, introduces compulsory arbitration, limits the closed shop, and penalizes labor for so-called unfair labor practices. It is directly contrary to the entire history and development of American labor.

Senator Ball, in trying to justify the introduction of this new Federal industrial relations bill, states in one of his articles published in the *World-Telegram* of July 9, "Neither labor nor management was consulted before the bill was introduced. It is obvious that the two partisans in this field are not ready at this time to agree and the authors of the bill are familiar with their positions generally."

That labor was not consulted—in this we agree with Senator Ball. Whether management was consulted—let us examine the members of the study committee who are responsible for this so-called "Federal industrial relations bill":

Charles B. Rugg, corporation lawyer of Boston, Massachusetts. He contributed five hundred dollars to the Republican Party campaign fund in 1944.

Kirk Smith, corporation lawyer of Providence, Rhode Island, a director of Hight St. Bank and Trust Co., also of four laundry companies.

Arthur Whiteside, president of Dun and Bradstreet, president of the Wool Institute; former president of the National Credit Office, etc.

George W. Alger, corporation lawyer of New York; counsel for Sheffield Farms, one of the big dairy concerns.

Harold G. Evans, president of the American Casualty Co., Reading, Pennsylvania, of the Acco Realty Co., and of the American Aviation and General Insurance Co., vice-president of the Reading, Pennsylvania, Chamber of Commerce.

Samuel Fels, president of Fels and Co., president of the Paschall Oxygen Co.

Donald R. Richberg, corporation lawyer, representing Standard Oil interests and other big businesses.

This committee did not have to consult management because it actually consists of management or representatives of management. This committee, representing large corporations, took it upon itself to "reform" the Wagner Labor Relations Act by the proposal submitted by these three gentlemen of the Senate under the guise of affording equality of protection to employers and employees alike. This bill practically eliminated most of the rights now guaranteed to American workers by the Wagner Act. By requiring a 75 per cent vote in order to win and maintain a closed shop, it abolishes the American system of deciding by a majority vote. Every fair-minded American will agree with labor that the elimination of a majority vote and the substitution of a three-fourth vote is, in itself, un-American, undemocratic, and contrary to the very principle of American democracy.

All sections of labor are united in opposing this bill. In

this transition from war to peace, while labor is striving to achieve industrial peace and to work with management and government in creating full production and full employment, an attack is being made on the basic rights of labor. I am confident that the achievement of American labor in the wartime period and the good sense of the American people will defeat any attempt against the well-being of all the American workers.

I have tried as briefly as possible to outline the role of American labor in the United States economy. Let me emphasize in closing that organized labor, now come to full maturity, intends to make its full contribution to preserving and extending full political and economic democracy in America. It extends its full support and loyal co-operation to all other progressive groups which are taking the same road. I know that if we work together, we can build a new life in this great and beautiful land of ours—a life from which poverty will be banished and in which men can live in security, peace, and plenty.

AMERICAN BUSINESS
AFTER THE WAR

*

BY WALTER D. FULLER

PRESIDENT OF THE CURTIS PUBLISHING COMPANY
PHILADELPHIA, PENNSYLVANIA

WHEN the guns of Europe ceased firing in May, American business was faced with the actual problem to which it had been giving careful—I might fairly say, prayerful—consideration almost since the beginning of the war. Back of the all-out mobilization of industry for war purposes loomed always the knowledge that some day —at a time unknown and practically unpredictable— American business would suddenly be confronted with the great problem of converting from war production back to peacetime production.

Winning the war came first; there was neither time nor inclination for anything else. Yet somehow after war councils were over, businessmen were able to reach conclusions which crystallized into a program of action. A year ago our plans were under way for a reconversion congress of industry to be held last December, but grave changes in the European situation led the government very properly to advise us to sidetrack reconversion for the time being and keep concentrated on war production.

RECONVERSION IS ONLY A TEMPORARY PROBLEM

We know that the end of fighting in Europe has not ended the war. But it did lessen the tension; it did give us a chance to look ahead to postwar problems as vital to the future of every American as arms and munitions are today. It even made some of those problems easier, for it appeared that we might do some reconversion while still engaged in major war production. And we needed to get our hands in again. After five years of disuse, our approach to peacetime requirements was a bit rusty.

But despite its crucial importance as a turning point in our economic life, reconversion is only a temporary problem. American business must go on, not only throughout the period of transition from war to peace, but for all the years of our nation's future. American business will do its share in adding brilliancy to that future.

LONG VIEW ESSENTIAL

Too often we take the short view instead of the long view. Few of you young people to whom I am talking know much of the great depression of the 1930's except as a matter of history. We survived it and passed into a great—though by no means happy—period of wartime achievements. To those of my generation who had to fight that depression it was terrible in all its impacts and implications. To many its barriers seemed unsurmountable.

Yet today it is so far behind us that our chief concern is that we should learn by its lesson. And, my friends, learn we must, for potential depression is a skeleton in the closet that is always with us. If we are smart enough and ingenious enough, we can keep the closet door closed, but the skeleton is forever there.

The pattern of economic cycles seems, if we are to judge by history alone, to be set in a repetitive mold. But if ever our study of history is to have value, it is at a time when historic mistakes can be rectified and when our knowledge of the past exposes disrupting influences which may affect our future. With war-bred stamina, courage, and united purpose, we can now pull together to achieve great peacetime goals as effectively as we have turned our united strength against the evils of tyranny and totalitarianism.

PRODUCTION WINS IN PEACE AS WELL AS IN WAR

There are interesting parallels between our present war prosperity and a prosperity we can have in the midst of peace. Even the terms are the same. Without production—and I say this humbly and not by one iota detracting from the splendid valor of our fighting forces—without production at home the efforts of all these brave men could have come to naught. And, again, without "distribution" boldly conceived and boldly carried through by sailors on the seven seas and by the men of the Air Transport Command production itself would have been useless.

Do these terms, production and distribution, have a familiar sound to you? Are they terms of both war and peace? Happily the answer is "Yes."

Production and distribution of goods and services are the essentials of peacetime prosperity just as production and distribution of guns and trucks and supplies have been the essentials of successful war. The difference is that in peacetime both production and distribution, to be successful, must be free from undue control.

99

PROSPERITY DEPENDS ON PRODUCTION

We have had some experience with governmental efforts to control America's economy. The idea of plowing under cotton and destroying little pigs seems ridiculous now, but it was no joke when it was going on. We tried to cure lack of balance in distribution by cutting supply; we "cut off our noses to spite our faces"; we tried to control prices; and the effort was scarcely effective within its limited aim. The broader implications of these misguided efforts were tragic; yet there are still people who believe that a controlled scarcity is worthy of trial.

Let us never forget that prosperity depends on the production, distribution, and use of goods and services, with profit to all. Please note that I include "use" as one of the bases of profit, for if there is no profit to the user, either in economy or satisfaction, he will not long remain a user, and both production and distribution will automatically cease. And so will jobs and the higher standard of living which we all desire.

CUSTOMERS MAKE JOBS

Those who look to postwar production as the provider of jobs—and that is as far as many people go—are putting the cart before the horse. Nobody can make jobs but customers. If customers want products and are taught to demand them, business can and will make and provide them. And in the making of new products many jobs will be created. Consumer demand is the real key to employment. Always the consumer is king.

Consumer demand creates payrolls, and those payrolls

are the basis for additional demands. Expanding demand, expanding production, expanding distribution—together these are the things that make more jobs. These are the things that make prosperity.

But prosperity includes more than the manufacturing industry. It includes a self-supporting agriculture, an efficient working force with high earning capacity, a vigorous system of distribution, and service for the products of both manufacturing and agriculture. Dollars flowing through these channels are the source of the prosperity we seek. Their rate of flow will measure the advance in living standards toward which we continually work.

PRODUCTS FOR THE FUTURE

While factories have been building tools of war, while laboratories have beeen distilling the essences of victory, there has been no brake on imagination. Even with military success as our most immediate and pressing task, we did not—we dared not—forget the peaceful future which has ever been our goal. From these wandering thoughts has come a new galaxy of hardheaded though fairylike dreams.

Let us see what research scientists have perfected in the laboratory and turned over to industry for development so that we consumers can all live more economically, more comfortably, and more luxuriously.

Chemists have developed liquid fertilizers from new sources that will have extraordinary effects on fruits and vegetables. The growth rate of potato seedlings has been increased 100 per cent by the use of new gases, which are also being used for ripening oranges. Aviation gas will be

101

broken down to stimulate the growth of fruit trees where the growing season is short. A yellow powder can increase the growth of fruits and vegetables to double their present sizes. New techniques in quick freezing and in dehydration will permit the conservation and easy distribution of constantly increasing varieties of perishable foods.

We know that there will be a tremendous need for new housing, in which many new materials will be used. Perhaps ready-made houses can successfully be turned out in the factory. Perhaps even your present houses can have many conveniences to be offered by the prefabricators. We will have hard-surface paints, repellent to dust. Light sources can be built into walls, ceilings, and furniture. Walls will tend to be made of materials which conserve either heat or cold. Consumers may have a variety of products made from light alloys and plastics.

And clothing! The silkworm, the sheep, and the cotton plant will have competition from rayon, nylon, rubber, spun glass, and the soybean. Consumers may be wearing milk silk or petroleum satin. Their clothes may be fireproof, mothproof, and even spotproof. We hope dresses won't wrinkle and pants won't shine and stockings won't run. Shoes may never be dull or leaky.

The postwar consumer's "platter party" may be supplanted by entertainment featuring records on a spool of wire which can be played a hundred thousand times with little loss of fidelity. In designing new cars, safety and visibility will be prime factors. Improved rubber may put a hundred thousand miles into tires.

The railroads are not lagging. Turbines and turboelectric drives plus modern coaches and better sleeping accommodations will give added comfort with greater speed and safety on streamlined trains. Noise and rough riding will be things of the past. Other forms of transportation—notably the commercial air lines—are all busily engaged in making plans for better public service.

MORE EMPLOYMENT THROUGH MORE OPPORTUNITIES

As living standards for the nation can only be raised and kept high by increasing production, so can more production be assured by expanding opportunities and the fullest freedom in taking advantage of those opportunities.

Americans who are enjoying the fruits of our advancing civilization aim to live well, to develop intellectually and culturally in an environment of comfort and convenience, and to preserve that vitality of mind, body, and spirit which has for generations kept America and Americans in the forefront of world enlightenment.

To realize these aims it is needful that there be opportunities for those who wish to engage in productive effort to reap reward in the degree their abilities permit. There must be jobs for those who want work and are able to work. This does not mean jobs *only* for industrial workers. It means jobs for farmers, jobs for distributors, jobs for the self-employed workers, jobs for people in every walk of life, and, also, *jobs for the savings of all these people.*

If we cannot find means to put our savings profitably to work, there can be little hope for the expansion which is needed to provide other kinds of jobs, particularly for all the young men and women who will be, year by year, entering the business world for the first time. Each of those

103

young people should seek out the place where he can best use his talents, and naturally each will be deeply interested in finding the economic climate which will most quickly ripen his harvest of achievement.

BUSINESS IS PREPARED

American business can produce. Its war record is little short of miraculous. With its sights set to produce eighty-five million pounds of airplanes in 1941, it made well over a thousand million pounds in 1944. Total war production by American industries was almost eight times as much in 1944 as it was in 1941. We produced a little over a million tons of merchant shipping in 1941, but in 1944 we produced eighty-one million tons. And we are still going strong. These are only samples.

The tools and factories are here, the men are here, the know-how is here; but all these things will not automatically turn themselves to peacetime production. The idea of turning "swords into plowshares" can be expressed in few words, but what headaches, what planning, what problems are there!

And, yet, with controls reasonably lifted, American business is ready for reconversion. Each American businessman, knowing that reconversion is *his* problem, has made his own plans, has prepared his own plant for the great change-over. The degree of preparedness is remarkable. We may have been unprepared for war, but nobody can say that American business is now unprepared for peace!

A recent survey made by the National Association of Manufacturers, covering 1,756 factories, shows that 61 per cent could resume peacetime production without delay and an additional 28 per cent will be changed over and producing within four weeks, leaving only 11 per cent which will require more than four weeks for reconversion.

These are heartening figures, for they reveal not only that the manufacturing industry clearly foresaw and prepared for the problems of reconversion, but that they are now ready for instant action when the green light comes!

IT TAKES BOTH IMAGINATION AND CASH

It seems self-evident that this preparedness, this readiness for immediate peacetime production, requires both daring imagination and hard cash. Industrial changes, be they large or small, cannot be made without expense; and it is a notable achievement that business management, in spite of a staggeringly high tax burden, has been able to conserve sufficient money to meet, in large part, the needs of reconversion. The financing of expansion, however, is another matter, with which I will deal later. If the half million new faces that enter business every year from our schools and colleges are to have their fair opportunity, our business structure must be expanded so that they may be comfortably housed without undue crowding.

As for daring imagination—when Uncle Sam has been your single customer for five years, where are you going to find the customers to take his place when he stops buying? Suppose you have the plant, you have the equipment, you have the raw materials, you have the men and women to fabricate them, but you have not, thank heaven, the warehouse space unendingly to store your products. Where will you find the users of your product? Creating demand is just as much a part of the producer's problem as is mak-

105

ing the goods. When you become a manufacturer you add a new word to your vocabulary, sales resistance. You will learn that people have to learn the usefulness of an unfamiliar article before they want it as their own.

We have only to go back to the beginnings of the automobile, which was a luxury for decades before it became the stand-by of every family. It took years to develop radio, even though every small boy struggled with his own set before his parents thought they could afford the "store-bought" kind. The early growth of the telephone was slow in the face of derision. When the phonograph was invented, its present uses were not even dreamed of. Its future was expected to lie principally in the realm of the office as a dictating machine! While the electric refrigerator was being perfected, it was no more than a rich man's toy, and there was little hope it would ever be more.

But in each of these cases imagination took over. Someone refused to believe in limited markets. Someone said that *these are things people want and need.* In each case imagination and courage created a market. The results are little short of fantastic.

EXPERT DISTRIBUTION REDUCES PRICES

As demands were created by imaginative distribution, quality improved and price went down. Between 1926 and 1941 the sales of electric refrigerators increased 579 per cent while the price went down 54 per cent; table model radio sales increased 135 per cent while the price dropped 79 per cent; silk stockings and electric irons cost 50 per cent less; cameras and toilet soap cost 60 per cent less. There are countless other examples.

Then came the war. And all these vast accomplishments, each of which was sparked by the individual initiative of freely competitive business organizations, were brought to a temporary halt. Our whole nation was mobilized behind the military forces, bending every effort to bring the quickest victory our organized resources would permit.

Soon we are going to be on our own again. Little by little the needs of our civilian population are being met. Bit by bit the enterprising businessmen of this country are redirecting their energies to picking up where they left off in 1941. None knows better than they that a successful and prosperous economy must be built up plant by plant and town by town. Prosperity cannot be handed down from Washington or from the Statehouse. It must be built up in thousands of localities by the hands and brains of people who transform our natural resources into all those things which give happiness and comfort to civilized men.

OPPORTUNITIES IN DISTRIBUTION

Never in history has so much thought and research been devoted to the development of new things. There is no dearth of things to make or of equipment with which to start making them. The one unpredictable factor is, How will the public take to all these fascinating novelties? Above all, how *much* and how *many* will it buy?

There are many critics of our distributive efforts. Some say that distribution takes too much of the consumer's dollar, yet they are the very ones who complain when they cannot find a package of cigarettes around the corner. Distribution is the most important of all our economic processes. You will probably be surprised—most people are—to learn that over half our working population is engaged in distribution and servicing things already made. And even with this large segment of our population already engaged in it, I believe that distribution offers greater opportunities to young men and women than any other phase of economic endeavor.

There is great hope for further improvement in distribution methods, yet I cannot believe we have done too badly up to date. Consider the spoonful of ground coffee that made your breakfast cup this morning: grown in South America, picked grain by grain, dried, shelled, packed, and assembled into shiploads of five to ten thousand tons per cargo, graded, blended, roasted, subdivided, packaged, wholesaled, retailed, and finally delivered to your cup at a cost of less than a cent—I think our distributive system, even in its crippled wartime shape, is worthy of our admiration.

BUSINESS IS A TURNING WHEEL

Recognizing, as we must, that distribution is already our greatest potential reservoir of employment and will continue to be so, it follows that increased efficiency in distribution is the greatest stimulant for production. A truly successful and growing economy calls for the orderly development of products new and old, both as to manufacturing them and making them easily available. Skill must be married to common sense.

As this orderly progress successfully takes place, there will be a demand for expansion of the means of production, additional working space and new machinery. Obviously such space and machinery cannot be obtained without money. And the two main sources of money for such use are past profits which have been retained in the business or investments coming from the personal savings of the American public.

The function of all business which gives real productive employment is to provide the right goods and services in the right proportions and at the right prices. Businesses are neither established nor enlarged unless there is a chance for profit—profit *both* to the consumer and to the maker. Profit to one alone is not enough. It is *both* these profits which permits, and encourages, growth. Without profit to *both* there can be no growth, no prosperity, and no new jobs. It takes a favorable business climate to bear fruit in American Homes.

NEW CAPITAL NEEDED

Profit is the product of skill. Two identical businesses can start side by side. One will succeed because it is skillfully managed; the other will fail from lack of skill. There must be skill in keeping down costs and skill in finding markets, skill in creating demands for goods, skill in adjusting production to distribution. The sum of these skills is translated into profit. Capital is useless without skill in its employment, but it is equally true that skills are useless without capital. Skills make not only standards of production and value, they make standards of use which give us ever-growing desires for higher standards of living in every home.

The growth of our economic system requires a continuous flow of new capital into productive channels. If that new capital is not forthcoming, production cannot expand. We have only to go back a couple of decades to see what happens to new capital when opportunity is stifled. Between 1920 and 1930, a decade of prosperity, the voluntary flow of new capital into American business averaged two billion dollars a year; but when the depression came and the chance for profit was highly restricted, new investment for the decade 1930–1940 dropped to an average of only half a billion dollars a year.

We are now faced with the serious danger that the obstacles to profit will again discourage the flow of capital, carrying in their train all the menaces to new products and new jobs. Capital, of course, only achieves value when it is turned into productive equipment—buildings, factories, tools, transportation, communication, and such dynamic implements. That is what businessmen have done with such savings as have survived our war taxes.

National wealth, which comprises our working resources, consists of property, goods, and services. Increased production of goods and services creates competition, not only to obtain customers and make profitable use of the fruits of industry, but also to obtain investors and make profitable use of their funds in providing the equipment needed for expanding production. It is axiomatic that those who succeed in best serving the public deserve the best reward.

SAVINGS ARE ESSENTIAL TO EXPANSION

Many enterprises, if they are to expand in accordance with public demands, will require the use of the savings of others. New enterprises, many of which will be organized,

110

will make use of large sums of venture capital. It is this money—venture capital, money risked in hope of fair reward—that makes new jobs. Venture capital, in our system of freely competitive enterprise, implies the risk of loss as well as the hope of gain. If the risk is great, the chance of profit must be proportionate to that risk.

It is easy to see how heavy taxation can be an obstacle to expansion. The taxes of industrious persons support the government. Whatever money the government uses is first earned by productive effort and is then removed by taxes from the realm of production. The more production there is, the more lightly may taxes be spread and the less the burden on each person. If encouragement of ability to pay taxes is to be a governmental policy, any revision of taxes should be so made as to bear easily on production, distribution, and employment. The more of each of these there is, the less oppressive will be the impact of taxation on everyone.

Every business executive is in effect a trustee. He works for three bosses—the consumers of his goods, his employees, and his stockholders. If he does a good job for the first two bosses, his results for the third boss are likely to be good also. But the three are independent legs of the same stool, and if one collapses, the whole stool falls.

FREE COMPETITION STIMULATES ENTERPRISE

Businessmen, as producers and distributors of the nation's available goods and services, are expected to maintain the competitive system which, as the basis of free enterprise, encourages every person to choose his own line of endeavor and through it prove to the world the value of his individual ability to aid our economic progress.

This competitive system implies, and in practice requires, that businessmen continually produce and distribute more and better goods at less cost; that manufacturers aid distributors in lowering costs; that businessmen set prices, in accordance with cost, which allow as fair rewards to workers and investors as they do to consumers; that they continue their valiant efforts to stabilize annual employment as a practical means of lowering unit costs; and that they stimulate economical production and efficient distribution through fair incentives to both workers and investors.

BUSINESS OFFERS INCENTIVES

Stripped of all folderol, of which I am sorry to say there is a great deal, jobs consist of opportunities for the worker to earn more money in somebody's employ than he can earn by his own unaided efforts. If business does not supply the incentive of good pay, plus the added incentive of still better pay for better work, it is clear that the worker will seek his living elsewhere. The same thing is true of savings. Business management must offer sufficient earnings on savings to make it worth while for the investors to risk this capital. When these conditions are met, there is an expanding economy.

REDUCTION OF GOVERNMENT DEBT

The gradual but steady reduction of the government debt, by repayment to our citizens of what is owed them, will make available large sums which can be used to expand our long-range peacetime economy. Government debt is not productive. It is, on the contrary, a withdrawal of funds from the economic stream. The same fact is one reason why high taxes have an unfavorable effect on our productive effort. Money paid out as taxes is not available for economic expansion.

On the other hand, industrial debt is contracted for productive purposes. Money so borrowed is invested in plant and machinery which create jobs. As good jobs increase, more payroll money is available to buy the products of other workers. This ensures a soundly expanding economy.

GOVERNMENT PROTECTS OUR FREEDOMS

Government is a necessary and an integral part of our social and economic organization and, like other elements of the organization, plays its role best when it is part of a balanced picture. Government is essentially protective, though it has functions beyond mere protection. While we want a government of laws rather than a government of men, we cannot overlook the tendency, so aptly stated, that "laws freeze our ambitions into patterns of yesterday."

Government has the positive function of protecting us in our freedoms, the freedoms of the Bill of Rights and the all-inclusive freedom of opportunity. Such restraints as are desirable are those which promote fair play for all and deny monopolistic privileges to any individual or group.

Free enterprise is vitalized freedom of opportunity—opportunity put to imaginative and constructive use. Every citizen should expect the government to protect his right to opportunity and his right to make the best use of it so long as it does not interfere with the opportunities of others.

In view of the war record made by industry, America looks to business management for leadership in promoting the country's long-range prosperity, though it requires teamwork from all. The government can be either highly helpful or dangerously obstructive. Government must regulate but never control. Through control comes the "national economic state." Along that road comes totalitarianism and eventually Nazism. We want none of this in America.

WHAT GOVERNMENT CAN DO

The government cannot further fair play and equal opportunity for all if it permits any of its subdivisions to go into the field of business enterprise. Judicial impartiality in regulation cannot be maintained if the regulator is engaged in any of the businesses subject to regulation.

Government can make its regulation protective rather than punitive. Naturally I do not mean there should be no punishment for wrongdoing. But I do mean that regulation which is so exacting as to discourage the expansion of any proper enterprise is socially undesirable.

Government can move toward the orderly reduction of the huge debt, which will be the chief cause of continuing high taxes, and can also move toward a reasonable reduction of all taxes as soon as the demands of war are satisfied. As a part of its tax program, it can bring about a speedy reduction in the cost of government itself. Most particularly can it cut out the weedy bureaucracy, which has all but choked the normal growth of proper regulatory functions. Government must be frugal if it expects its citizens to be so.

Government can be a powerful factor in promoting good industrial relations by making all parties in collective bargaining and other labor negotiations equally responsible. Government should take all needed steps to stamp out monopoly wherever it occurs—in business, in labor, or even in government itself. American business believes in free competition and is emphatically opposed to monopoly or to cartels in any form.

The prosperity we all want can best come about if the government takes stern steps to stabilize the currency and adopts a forward-looking policy which will expand foreign trade so that all will benefit.

LONG-RANGE GOALS FOR EMPLOYMENT

The young men and women just entering business have the same practical needs, as well as the same hopes and ideals, that inspire all other workers who are a vital part of going industry. Satisfaction with both working and living conditions among employees is quite as important to management workers as to the individuals directly concerned.

Only through mutual respect and good feeling can we journey together toward the long-range goals which are the objectives of all engaged in business—management workers and production or distribution workers alike. We have already gone a long way toward reaching these goals. We all have made mistakes in the past, and perhaps it is too much to hope that the future will be without error. Of this I am sure, however. As we come to a better understanding of each other's needs and problems, the chance of error becomes progressively less.

The most important of our long-range goals—the one on which all others depend—is the preservation of the rights, freedoms, and opportunities which are ours under the American Constitution. Such preservation goes beyond strict interpretation to include the spirit which led to the enumeration of the rights on which freedom and opportunity rest. An understanding of these rights lays the groundwork for that mutual effort through which all can most quickly make a success both of work and of the finer living which is the reward of work.

There is the goal of pay and profits based on performance. There is the goal of steady employment, freely sought and freely found. There is the goal of professional rewards and a self-reliant agriculture. There is the goal of a chance for everybody to be happy, to be free, to save, and to put his savings to work in building a competency for self and family.

There is the goal of ever-higher standards of living. We do not need to look very far back to see what has been accomplished through enlightened business effort to create more leisure and to make the use of that leisure more pleasant. For example, statistics show that during the first thirty years of this century the horsepower of factory tools more than tripled and the volume of manufactured goods more than doubled. This provided work for 88 per cent more workers in industry, and leisure time for the average worker was increased by 27 hours per week.

The mere cataloguing of our goals shows clearly that what is best for the citizen and for the consumer is best for business itself. As I have said, the transition problems are important, but temporary. It is the long-range goals we must keep constantly in mind.

BUSINESS IS EVERYBODY AT WORK

It takes no prophet to point out that only through all branches of capital, labor, inventors, consumers, and management growing together and working together can we achieve a better America. As businessmen have overcome obstacles in the past, so they are overcoming many at the present time. And there is no doubt that the obstacles of the future will also be met and, in many instances, be utilized to spur us on to even greater accomplishment. Tough going hardens the muscles, both physical and mental; builds self-reliance; and gives savor to success that no easy way can equal.

Business is only the community earning its living. If that living is to be good, business must be good. Economic liberty is inseparable from all other liberties which we so rightfully cherish. The basic economic liberty, the one which needs to be most closely guarded, is freedom of opportunity. If free opportunity continues, then we have little to fear for the coming years.

PLANS FOR THE FUTURE

After the last war we had our problems of postwar recovery. At that time the vast productive resources of business, which had been prepared for war, were put to work on the products of peace. In the forefront were the automobile, the radio, and a multitude of household electric appliances. The public was ready for improvement in

117

its living standards, and the problems of distribution were handled brilliantly.

Today we have far greater productive resources, and we must direct our attention still more directly to distribution, marketing, and to the service industries.

A survey made by Dun & Bradstreet, business statisticians, for the fourth quarter of 1944 showed that, of over twenty-two thousand business institutions answering their queries, 37 per cent of the manufacturers and 44 per cent of the wholesalers contemplated adding new lines or products to their businesses. Note that the wholesalers, who are basic distributors, contemplated even greater expansion of lines than did the manufacturers.

This survey also showed that the wholesalers led in plans for new classes of customers. Thirty-four per cent of the wholesalers expected such a type of expansion, 46 per cent expected to expand their sales territories, and 27 per cent planned new distribution methods or channels.

The American Legion estimates that 40 per cent of all wartime displaced workers will find their opportunity in distribution as against about 25 per cent in manufacture.

Training in distribution and its problems opens real opportunity both for educators and for students interested in a bright future. Advertising, selling, transportation, markets, consumer needs and wants are all areas for research which can often develop unexpected demands, savings, profits, and social advances.

It is within the province of business to see that latent purchasing power is translated into a steady and vigorous demand for more useful goods and services, when supply

118

and demand for civilian goods once again approaches equilibrium.

IMPROVED HEALTH AND SANITATION

More comfortable homes, more nutritious and easily prepared foods, swifter and more flexible transportation and communication, and better educational and cultural opportunities, brought about through business enterprise, have made our standard of living the highest in the world. The importance of health and sanitation is constantly being more widely recognized. Here is the opportunity for expanding an industry, of which the modern bathroom is only the beginning.

Public health boards are setting up more vigorous standards of sanitation in restaurants and other gathering places. Meeting these standards will require new equipment, to a greater or lesser degree, in every establishment affected. Public water supply and waste disposal will be required to meet standards of purity and sanitation which would have seemed like mere fantasy a few years ago, but which are becoming imperative as all the factors that increase the span of human life are better understood.

Still better foodstuffs, still more comfortable and health-promoting beds and chairs, still more functional clothing, still greater strides in the realm of protective and auxiliary medicine—all these mark the trend of future economic advance. They mark the trend of future opportunity to expand business on the sure and solid foundation of giving humanity something in which humanity's profit is greater than any which can be collected by the enterpriser.

119

NEW BUYING HABITS

The increased human life span which has already come through well-directed efforts in health promotion and sanitation will have a marked effect on the buying habits of people. There will be an increasing tendency to develop articles, both medical and domestic, suitable for older people. As the average age of our population increases, with more people past the age of retirement, means will be developed to give them more cultural opportunities as well as more detailed facilities for comfortable older years.

As business has constantly, through improved processes, helped workers to earn more in shorter hours, there has come about an expansion in travel habits which is not far past a good beginning. Air travel, a great business itself, will be parent to a multitude of auxiliary businesses, just as we have seen the automobile business spawn a myriad of related industries. And the railroad and ships and the automobile offer great opportunities born out of wartime discoveries and necessities.

The expansion and diversification of travel will bring about many changes in shopping practices, each of which will be studied by alert distributors anxious to increase the range of their services.

HOME IMPROVEMENT

During the past two decades considerable progress has been made in the development and use of new materials in home construction. New construction features have also been developed which add to home comfort and livability. This particular period has also witnessed wide growth and

120

public acceptance of many mechanical household appliances and items of labor-saving equipment, which have resulted in easing the burden of domestic chores for the American housewife.

Progress in these fields has not yet reached its limit and seems capable of much further expansion and probably a more rapid increase in the future. Many of these developments, which were formerly luxury or semiluxury items, are now considered to be home necessities, and almost every survey of consumer intent indicates a tremendous pent-up demand for these types of goods and services.

Improvements in design and changes in style may increase the rate of obsolescence on many items of equipment and appliances and can be expected to accelerate their rate of replacement. This will necessitate the development of markets for used appliances and equipment, bringing these mechanized conveniences within the purchase range of many families who could not afford them in the past. The net result of this should be to make modern conveniences available to a gradually broadening market at relatively lower and lower costs. All this holds forth considerable promise to those industries which cater to the home.

Accompanying this growth in markets will be the need for more and better servicing facilities, combining vast opportunities for employment with a profitable business function. As a result of their war experience, many businesses, particularly those in the household appliance field, have found that the servicing and repair of these items, instead of being a nuisance or a necessary evil, can be turned into a builder of good will and made to yield a handsome return. It is to be expected, therefore, that servicing of products will continue to grow and occupy a more prominent part as one of the essential functions of distribution. It seems likely that the bulk of this activity will be performed at the retail level and will require specialized knowledge and skills covering a wide variety of products.

SKILL, ORIGINALITY, INITIATIVE

There will be many new businesses after the war—new in organization as well as new in kind—and the survival of many of these, faced as they will be by the keenest kind of competition, will depend on the skill, originality, and initiative of their distributive organizations.

The economic trends of the future will be charted by research, but, as always, they will be influenced by the initiative of individuals. None of the great business enterprises of which Americans are proud has failed to bear the imprint of one or more outstanding individuals who have, by coupling imagination with resourcefulness, made many more than the traditional two blades of grass grow where one had grown before. Many of these men are still hustling in business, and their example inspires the young people of today to search for future opportunities with the same self-reliant will which has always demonstrated leadership.

Both agricultural and manufacturing production may be expected to work more closely with distributive outlets to bring about attractive prices which will encourage consumers to make wise use of their buying power. Based on the known fact that consumers insist on buying where they can get the most for the least, there is no doubt that closer co-operation between producer and distributor will give us constantly better goods at lower and lower prices.

BUSINESS IS ALWAYS YOUTHFUL

Business is prepared for the future. Individual managements throughout the country have studied their own opportunities for postwar expansion and are working wholeheartedly toward achieving a lasting prosperity for America. A recent survey shows that plans for new products to be made by existing plants will call for an increase in employment in manufacturing alone of over 30 per cent above the 1939 figure. This does not include the hundreds of thousands of new enterprises which will be launched.

Under the American economic system, which has continuously advanced within the framework of free opportunity for free enterprise, there has always been assurance that youth and the youthful outlook would be the best guarantee of rapid, substantial, and imaginative progress. I am sure that coming generations of American young men and women will make free use of their opportunities.

Free opportunity needs sound thinking, sound planning toward sound objectives, self-discipline, and self-restraint to the end that others' opportunities shall not be impaired and that success may be attained. It implies a self-inspired spirit of competition to create new economic outlets. It implies a diffusion of competence through the country which will in itself protect all against restrictive concentrations of power in any form. Successful free enterprise calls for the active application of the Golden Rule—first, last, and all the time.

Free opportunity does not come of itself. The efforts of

business are now, and should always be, directed toward the maintenance of high employment, through which purchasing power, the mainspring of prosperity for all, can be maintained.

Freedom of opportunity carries obligations with it. It calls for competency in citizenship—a competency which is backed up by study and knowledge of the responsibilities each person has toward all the other persons in his community and country. It calls for competent consumers. It calls for competent and creative workers, contributing those individual drops of effort which combine to make the ever-increasing flow of constructive accomplishment that makes this country great.

POWER POLITICS AND INTER-NATIONAL ORGANIZATION

*

BY HERBERT W. BRIGGS

PROFESSOR OF GOVERNMENT,
CORNELL UNIVERSITY

THE CHARTER of the United Nations signed at San Francisco on June 26, 1945, states in Article 2 (1) that "the Organization is based on the principle of the sovereign equality of all its Members." The flaunting of these two concepts of the sovereignty and the legal equality of states, neither of which was expressly referred to in the Covenant of the League of Nations, might seem unfortunate were not both concepts flouted in numerous other provisions of the Charter. The outstanding characteristics of the Charter are its recognition of the actual and legal inequality of the Members of the United Nations, and its provisions empowering the Organization to take action, binding on its Members, without their unanimous consent. Since these features dominate the structure of the Organization, I should like to direct your attention first to the basic theories and conditions of fact which have dictated their choice.

From the bench of the United States Supreme Court, Justice Oliver Wendell Holmes once made the pronouncement that "sovereignty is pure fact"; but since, in a later case, he stated that "a word is not a crystal, transparent and unchanged; it is the skin of a living thought and may vary greatly in color and content, according to the circumstances and the time in which it is used," we may perhaps be pardoned for asserting that sovereignty is not a fact, crystalline or otherwise: it is a juristic theory. The context of the juristic theory was originally the France of the sixteenth century in which it seemed desirable to Jean Bodin to support the authority of the king over the Church and the nobility as the sole source of law, although even kings were expected to observe treaties. Viewed internally, the juristic concept of sovereignty may have performed a useful role in emphasizing that within each state there must be one, and only one, source of law in a formal sense. Yet, the transfer to international relations of a concept which emphasizes the supremacy of the state in the field of law has meant, in the words of one writer, that "the sovereign state does not acknowledge a central executive authority above itself; it does not recognize a legislator above itself; it owes no obedience to a judge above itself." Another writer ascribes to doctrines of sovereignty the function of laying "an impenetrable smoke screen round the key position in any system of power politics, the position occupied by the sovereign state."

In the courts, a more limited conception of external sovereignty has prevailed. Thus Judge Max Huber in the *Palmas Island* case before the Permanent Court of Arbitration stated: "Sovereignty in the relation between states signifies independence. Independence in regard to a portion of the globe is the right to exercise therein, to the exclusion of any other state, the functions of a state." Similarly Judge Dionisio Anzilotti in the *Austro-German Customs Regime* case before the Permanent Court of International Justice was of the opinion that external sovereignty or independence meant "that the state has over it no other authority than that of international law." Perhaps one should not quarrel with a conception of sovereignty as independence, subject to law; yet the form in which it is expressed soothingly misdirects our attention from the essential problem—the creation of institutions to establish the rule of law.

Akin to the doctrine of sovereignty in its practical consequences is another juristic concept: the legal equality of states. It is twenty-five years now since Edwin Dickinson first pointed out in his book, *The Equality of States in International Law,* that this conception, nurtured in the doctrines of natural law and natural rights and transferred by analogy from persons to states, has taken two divergent forms. The first, equality before the law, was expressed by Léon Bourgeois at the Second Hague Peace Conference as involving for each state, whether large or small, "an equal claim to respect for its rights, an equal obligation in the performance of its duties." The more extensive second form, equality of rights, was stated by Chief Justice John Marshall in the case of *The Antelope* as follows: "No principle of general law is more universally acknowledged, than the perfect equality of nations. Russia and Geneva have equal rights." And Sir William Scott, referring in *Le Louis* to "the perfect equality and entire independence of all distinct states" as a fundamental principle of public law, added: "Relative magnitude creates no distinction of right; relative imbecility, whether permanent or casual, gives no additional right to the more powerful neighbor. . . ."

The first of these meanings of state equality—equal protection of the law or equality before the law—is essential, writes Dickinson, to any advanced legal system, but "is not inconsistent with the grouping of states into classes, each of which the law regards differently." Thus each state has an equal claim to respect for whatever rights it may have and an equal obligation in the performance of whatever duties it may have; but to assert, as does the second conception of state equality, that all states have equal rights— or even an equal capacity for rights—is contrary to fact, and its advocacy has stunted the growth of international institutions of government. Let me illustrate the point by a reference to individual equality within the state. Within a political society the attainment of equality before the law has not involved as a necessary consequence an identical equality of rights among men; not all men are sheriffs, judges, or legislators. Yet in the international field each state, basing its demands on concepts of sovereignty and absolute juridical equality, has traditionally claimed a right of equal participation not only in quasi-legislative bodies but in the executive councils and judicial organs designed to apply the law.

It is the practical consequences drawn from these concepts with reference to international organization which justifies their discussion here. More specifically, the development of international institutions adequate to serve the needs of the international community has been thwarted by claims derived from the concept of sovereignty that no new rule of international law is binding on a state without its consent; that any membership in, or compulsion by, an international organization must be based on consent; that no court or council has jurisdiction over the acts or disputes of a state without its consent. Similarly derived from

130

the concept of equality have been claims of equality of representation, membership, and voting in international organizations and their organs; the unanimity rule (i.e., the rule that the majority does not bind the minority); and an equality in refusing to ratify decisions of international organs.

The attempt during the past century to establish these claims as principles of law has come up against the hard fact of the actual inequality of states. In the light of this fact, concepts of sovereignty and equality have played a dual role, supporting the Great Powers in their refusal to accept legal limitations on their actual power and being used as a largely ineffectual shield by the smaller states in an attempt to compensate their actual inequality. "The most impressive feature of the international society is clearly its hierarchic structure," writes Georg Schwarzenberger, the author of a book entitled *Power Politics*. The distinctions between World Powers, Great European or Asiatic Powers, intermediate states, and lesser states depend on the over-riding criteria of power. "In such an environment," he adds, "it becomes the supreme duty of any entity which wishes to survive to acquire at least as much power as is required to be able to carry on the perpetual struggle for existence."

With the criteria of power—military, economic, technological, geographical, human, and political—I am here less interested than with the politics of power, with purposes, methods, and results. Since the power of a state has meaning only in relation to the power of other states, there is a constant struggle for security, expressed in negative terms as the power not to be coerced. Writing in *Foreign Af-*

131

fairs in 1932, Karl Radek, then editor of *Izvestia,* said: ". . . the Soviet Union cannot remain indifferent to the changes which are taking place in Manchuria, through which or near which pass the lines of communication giving access to her Pacific ports. . . . The Soviet Union is strong enough to defend her territorial integrity and her interests. Concentrating her efforts on building up peaceful industries for meeting the needs of her own population, keeping aloof from armed interference with the affairs of foreign nations, the Soviet Union will seek a peaceful settlement of all conflicts which may arise between her and her neighbors. She will base her policy exclusively on her own interests, which correspond with the interests of peace both in the east and in Europe." Whether or not the expression of power in terms of security is merely the tribute which vice pays to virtue, each of us can decide for himself; but even if we admit that national security is the primary purpose of the politics of power, it remains true that in a system of power politics actual power not to be coerced involves the power to coerce. Poland is a neighbor of Soviet Russia; and security, like power, is relative.

The pursuit of national interests by the methods of power politics is based on what Schuman terms "the assumption of violence": a willingness to use the instruments of power, to take the risks involved, to use or threaten force in order to maintain or improve one's relative position. Moreover, an important characteristic of the methods of power politics is the reliance on national action. Intervention, military power, imperialism, even policies of alliance and balance of power are methods by which an individual state arbitrarily employs physical force or

132

more subtle pressures, politics, or realistic bargaining to secure its national interests. The methods may be contrasted with collective security—a system in which the threat or use of force in international relations would be controlled by a responsible organ of the international community in accordance with generally accepted principles.

That the pursuit of national security through the politics of power has sometimes resulted in a fortuitous world order cannot be denied. More or less precarious equilibria between the Great Powers have temporarily stabilized large areas; and the security of smaller states has sometimes rested on the fact that they were focal points of tension between the Great Powers. Where such equilibria and tensions were for a time established, they had the virtue of being in accord with the realities of power; but the vice lay in the essential instability and appalling irresponsibility of a system in which peace and international order were merely chance results of the national pursuit of power.

The need for a more direct attack on the problem was given some recognition in the nineteenth century. During periods of equilibrium among themselves, the Great Powers realistically adjusted tensions, reconciled divergent interests, liquidated trouble spots, preserved order among lesser states, and in general exercised executive functions of government. By the Treaty of Chaumont, signed in March, 1814, Austria, Russia, Prussia, and Britain agreed "to devote all the resources of their respective states to the vigorous prosecution of the present war"—against France —"and to employ them in perfect concert, in order to obtain for themselves and for Europe a general peace, under the protection of which the rights and liberties of all nations may be established and secured." At the Congress of Vienna, all Europe except the Turks was represented. The smaller states came expecting that the Congress would be a European parliamentary assembly in which they would participate on a basis of equality with the Great Powers; but the latter had no intention of permitting the small states to remake the map of Europe. Lord Castlereagh, British Foreign Minister, had prepared an elaborate scheme for convening the Congress in plenary session merely to approve decisions previously made by the Big Four. In this way, he argued, "you obtain a sort of sanction from them for what you are determined at all events to do." When Talleyrand, French Foreign Minister, arrived in Vienna, he championed the rights of small states and insisted on a full meeting of the Congress; but the Big Four admitted France, making it the Big Five, and the Congress never met in plenary session.

At Vienna, writes Dickinson, the Great Powers "rearranged the map of Europe, restored dynasties, confirmed the partition of Poland, united Belgium with Holland, neutralized Switzerland, created the German Confederation. . . ." As Lord Palmerston later observed: "The tide of war had swept over the whole surface of Europe from the Rhine to Moscow, and from Moscow back to the Seine; all the smaller States of Europe had been conquered and reconquered, and were considered almost at the arbitrary disposal of the Great Powers whose armies had decided the fate of the war. . . . The smaller Sovereigns, Princes, and States, had no representatives in the deciding congress, and no voice in the decisions by which their future destiny was determined. They were all obliged to yield to overruling power, and to submit to decisions which were the result, as the case might be, of justice or of expediency, of generosity or of partiality, of regard to the welfare of nations, or of concession to personal solicitations." By the Quadruple Alliance, the Great Powers agreed to renew their meetings to consider measures which "shall be considered the most salutary for the repose and prosperity of nations, and for the maintenance of the Peace in Europe"; and at Aix-la-Chapelle in 1818 it was made clear to small states claiming participation that no principle of the equality of states entitled them to admission. The Confederation of Europe soon degenerated into a league to maintain despots in power against the rising tides of democracy and nationalism, but as W. Alison Phillips has written, the significance of the Confederation "is, that it represented, whatever the motives of the several Allies may have been, an experiment in international government."

The Confederation of Europe was shortly replaced by the Concert of Powers, of which Drouyn de Lhuys was later to say: "It is the five powers to which belongs the right to regulate interests which affect Europe as a whole"; and M. Hanotaux: "The European Concert is the sole tribunal and authority before which everybody must bow." Particularly with reference to the Eastern Question, writes James W. Garner, the Concert assumed authority "to dictate settlements, establish arrangements, and to supervise their execution." It freed Greece, Roumania, Serbia, and Montenegro from Turkish rule; "in some cases their kings were selected with the approval of the Concert; their constitutions were submitted to its approval"; it drew their frontiers, and placed some of them under its guaranty.

135

Elsewhere, "it permitted the dissolution of the unnatural union between Belgium and Holland and forced the latter to accept its decision; it neutralized Belgium, Luxembourg, the Black Sea . . . ; it blockaded coasts to prevent hostilities; it sent troops to Syria to pacify disturbances there; it established a system of control over the finances of Egypt and Greece; it exercised collectively the power of coercion, restraint, legislation, supervision and guardianship over a considerable part of Europe." The sovereignty and equality of states, he adds, "found little recognition in the numerous conferences which were held to regulate these affairs or in the decisions which were reached. The right thus asserted and exercised by the Powers had no legal foundation and no political basis other than the claim that it was the right and duty of those which had the power to exercise a guardianship in the interest of the general peace and public order."

Sporadically throughout the nineteenth century the Great Powers, acting collectively, regulated certain affairs and certain areas. They succeeded only where there existed either a community of interests among themselves, or where opposing groups were sufficiently balanced in power to prevent unilateral action by one state or one group to the detriment of others. The interests and purposes of each of the Great Powers were basically national. For example, national security was served by stability in a certain area (the France of the post-Napoleonic era; the Balkans throughout the nineteenth century) or by controlled change (an assurance sought that the breakup of Turkey's European possessions would not give to any Power a preponderant advantage in southeast Europe or

136

the Mediterranean area). However, a more than incidental consequence of collective action by the Great Powers was a degree of stability and world order. To the extent that they adjusted inter-Power rivalries, intervened collectively in the affairs of smaller states, or liquidated tensions, the Great Powers filled a void: in a functional sense they constituted an international security council.

Their authority, however, was self-assumed, self-defined, and self-limited. Despite their occasional assertions that they were acting for the international community of states, there was no defined responsibility, no constitution, indeed, no organization of the community which could confer authority or entrust them with responsibility. Nor did the Concert itself have any permanent organ meeting periodically for expediting discussion or permitting prompt action—a lack which was deplored in the summer of 1914 when the Concert had long since split into rival camps and the lights were going out all over Europe.

In 1919, when the League of Nations was being fashioned in Paris, there were those who, cognizant of the lessons of the European Concert, wished to build the League around the idea of the preponderance of the Great Powers. Certainly the Peace Conference itself exemplified this preponderance. As at Vienna a century before, the Great Powers—this time the United States, the British Empire, France, Italy, and Japan—completely dominated the Paris Peace Conference, determining its organization and procedure, deciding that they alone could attend all sessions of the Conference and its commissions, though permitting representatives of smaller states to attend meetings at which questions concerning them were discussed, and making all important decisions, though permitting the smaller states to ratify the decisions of the Council of Ten or the Council of Four in plenary sessions of the Conference. In such an atmosphere it is not surprising that Lord Robert Cecil envisaged a Council of the League of Nations composed exclusively of the Great Powers, who should meet annually, while the principle of the equality of states would be relegated to an Assembly of all League members meeting every four years. He stated that the success of the League would depend upon the support of the Great Powers; they would run the League, and it was just as well to recognize it flatly as not; and the small states would probably join anyway.

The remarkable Smuts Plan, subtitled "A Practical Suggestion," employed a different approach to much the same end. Rejecting the idea of a superstate, General Smuts said: "But while we avoid the super-sovereign at the one end, we must be equally careful to avoid the mere ineffective debating society at the other end. . . . We want an instrument of government which, however much talk is put into it at the one end, will grind out decisions at the other end. . . . The league will include a few great Powers, a large number of medium or intermediate states, and a very large number of small states. If in the councils of the league they are all to count and vote as of equal value, the few Powers may be at the mercy of the great majority of small states. . . . The league is therefore in this dilemma, that if its votes have to be unanimous, the league will be unworkable . . ." ("nobody will take the league seriously"; "it will soon be dead and buried, leaving the world worse than it found it"); but if votes "are decided by a majority, the Great Powers will not enter it; and yet if they keep out of it they wreck the whole scheme." His proposed solution was to recognize the equality of states in the Assembly but to limit its power to mere discussion and recommendation on matters submitted to it by the Council. "The real work of the league" would be done by its Council, to which, along with the five Great Powers and, eventually, a democratic Germany, General Smuts was willing to admit a minority of lesser states, with any three or more states having power to veto action. Cecil opposed this concession to the smaller states, and it was the substance of the Cecil plan for a League dominated by a Council composed exclusively of the Great Powers which became the basis of discussion before the League of Nations Commission of the Peace Conference.

The Covenant which eventually emerged was a victory for the smaller states. Recognition of the preponderance of the Great Powers was limited to permanent membership in the Council; the Assembly was to have equal authority with the Council on "any matter within the sphere of action of the League or affecting the peace of the world," although the Council was given priority in certain matters. The sovereign equality of states, while not expressly mentioned in the Covenant, found embodiment in the form most feared by General Smuts: the unanimity rule giving to every state, large or small, an absolute veto over practically all action by the League.

Although sixty-three of the world's seventy-odd states were sometime Members of the League of Nations, at no time were all the Great Powers parties to the Covenant; and the withdrawal of Japan, Germany, and Italy, the

139

early abstention and later expulsion of the Union of Soviet Socialist Republics, as well as the continued abstention of the United States, were reflected in the power potential of the League. True, Maxim Litvinov declared in February, 1938, when, of the Great Powers, only Britain, France, and the Soviet Union were participating in League affairs: "There is no state or bloc of states which could resist the united forces of the Members of the League, even as it is composed today"; but, whether his estimate was right or wrong, the absence from Geneva of Germany, Italy, Japan, the United States, and almost a score of smaller states, undoubtedly conditioned the expression of League power.

Moreover, within the League, the composition of the Council never accurately reflected the power situation. The four or five Great Powers in the League at any one time were never a majority of the Council and were at times confronted with from nine to eleven small, and frequently very weak, states, the vote of each of which was equal to that of a Great Power, with all its population, its territory, and its industrial and military potential. The not unnatural result was an "inner circle tendency" in the Council. A few gentlemen, representing only the Great Powers, met over a cup of tea or more potent libations and made decisions, as at Locarno, in advance of League meetings, and subsequently steam-rollered the decisions through the League. A more effective method was to deprive League organs of jurisdiction entirely on the plea that the matter was already being dealt with, or could better be handled, by the Conference of Ambassadors in Paris, an organ of the Great Powers, which, unhampered by League principles, disposed of many matters with considerable realism between 1920 and 1931. To attribute the failure of the League to the small states would be to misconceive the problem: it was the unworkable nature of the doctrine of equal rights that caused power, like water, to find its level, and the real decisions to be made outside Geneva.

The lessons were not lost on the drafters of Dumbarton Oaks and San Francisco; and in their approach to the problem of building a security organization which might work with, rather than in opposition to, the realities of power, they gave legal and practical recognition to the preponderance of the Great Powers and largely denied to small states those consequences derived from theories of sovereignty and equality which might hinder the Organization from taking, in the somewhat less than beautiful words of the Charter, "effective collective measures for the prevention and removal of threats to the peace, and for the suppression of acts of aggression or other breaches of the peace," or from bringing about "by peaceful means, and in conformity with the principles of justice and international law, adjustment or settlement of international disputes or situations which might lead to a breach of the peace."

The preponderance of the Great Powers is recognized by the United Nations Charter in several ways. In addition to being members of the General Assembly on a basis of equality with all Members of the Organization, China, France, the Soviet Union, the United Kingdom, and the United States are permanent members of the Security Council, which will also include six other states elected by the General Assembly by a two-thirds vote. A more accurate reflection of the world power situation would have limited membership in the Security Council to the three real World Powers—the Soviet Union, the British Empire, and the United States; but the drafters wisely included a potential Great Asiatic Power, China, and a potential Continental European Great Power, France. Unlike the League of Nations Covenant, the United Nations Charter makes no provision for the naming of additional permanent members of the Security Council, so the chances of a local boy making good are definitely discouraged. The inclusion of six smaller states in the Security Council is counterbalanced by the second special privilege of the Big Five, an absolute veto on most matters—a veto which is not given to the smaller members. The third means by which the preponderance of the Great Powers is enhanced is found in the strictly limited powers conferred by the Charter on the General Assembly and organs other than the Security Council.

Another way of securing the special position of the larger states finds but scant recognition in the Charter. This would be the weighting of votes in the General Assembly, such as is found in other international organizations like the Bretton Woods International Monetary Fund and International Bank for Reconstruction and Development in which the number of votes a state possesses depends upon its quota or subscription, and the Universal Postal Union in which, in effect, extra votes are given to states with colonies. The granting of the demand of the Soviet Union for separate membership in the United Nations for two of her sixteen principal political subdivisions —the Ukrainian and Byelorussian Soviet Socialist Republics—, aside from rendering the phrases in the Charter about the sovereign equality of all Members even less meaningful, is the only present step in this direction.

The significance of Great Power preponderance in the Organization will require further treatment, but it should be noted here that this preponderance is coupled with something entirely lacking in the Concert of Europe— namely, the principle of delegated responsibility. Not only is the authority which is obliquely conferred on the Great Powers by the Charter a delegated authority, but its exercise is circumscribed by an explicit statement of the purposes and principles in accordance with which the Organization and all its Members, including the Big Five, shall act. The Purposes of the United Nations (Art. 1), briefly stated, are to maintain international peace and security; to develop friendly relations among nations; to take measures to strengthen universal peace; to achieve international co-operation in solving international problems of an economic, social, cultural, or humanitarian character; to promote and encourage respect for human rights and for fundamental freedoms for all; and to be a center for harmonizing the actions of nations in the attainment of these common ends.

The Principles in accordance with which the Organization and all its Members shall act (Art. 2) include first our old friend that "the Organization is based on the principle of the sovereign equality of all its Members." In the context of the entire Charter this must be taken to mean merely that the United Nations is not a superstate, but a confederation whose Members retain independence, subject to the legal obligations of the Charter. This interpretation is supported by the second Principle, according to which "all Members, in order to ensure to all of them the rights and benefits resulting from membership, shall fulfil in good faith the obligations assumed by them in accordance with the present Charter"; and by the seventh Principle which states, in part, that the Organization is not authorized "to intervene in matters which are essentially within the domestic jurisdiction of any state." This seventh Principle, which fails to provide for judicial determination of what constitutes an essentially domestic question, is a potentially dangerous limitation on the jurisdiction of the United Nations to accomplish some of its declared purposes, although a proviso adds, in effect, that a plea of domestic jurisdiction cannot be used to prevent enforcement measures by the Security Council in dealing with a threat to the peace or a breach of peace.

Principles 3 and 4 are exceedingly important, since, in connection with other provisions, their effect is to give a monopoly [1] on the legal use of force to the United Nations Organization. Principle 3 reads, in part: "All Members shall settle their international disputes by peaceful means . . ."; and Principle 4 states: "All Members shall refrain in their international relations from the threat or use of force against the territorial integrity or political independence of any state, or in any other manner inconsistent with the Purposes of the United Nations." These stipulations are no mere renunciation of war, but couple a positive obligation to settle international disputes by peaceful means with a prohibition on the threat or use of force.

---

[1] With the exception of a limited right of self-defence under Article 51 and the right to take certain enforcement measures against enemy states of the second World War under Articles 53 and 107.

Noteworthy is the omission of the proviso found in the League of Nations Covenant that a state might lawfully resort to war three months after a failure to settle a dispute by certain pacific means.

Principle 5 of the United Nations Charter establishes in limited form the principle of collective security, by stating: "All Members shall give the United Nations every assistance in any action it takes in accordance with the present Charter, and shall refrain from giving assistance to any state against which the United Nations is taking preventive or enforcement action." I say "in limited form" because any one of the Great Powers can veto preventive or enforcement action by the Organization and to that extent the security of a state is not collectively guaranteed.

The sixth Principle of Article 2 of the Charter is curious, reading: "The Organization shall ensure that states which are not Members . . . act in accordance with these Principles so far as may be necessary for the maintenance of international peace and security." This means that the Organization shall use force, if necessary, to compel a non-Member to act in accordance with principles by which it is not legally bound. It also raises the whole important question of membership in the Organization.

Article 4 of the Charter, after providing that "Membership in the United Nations is open to all . . . peace-loving states which accept the obligations contained in the present Charter and, in the judgment of the Organization, are able and willing to carry out these obligations," adds that the admission of any such state as a new Member shall be by a two-thirds vote of the General Assembly upon recommendation of the Security Council. These provisions

145

contain several unfortunate features. Since such a recommendation of the Security Council must include the votes of each of the Big Five, any one of them could veto the admission of, for example, Switzerland or Sweden—just as Woodrow Wilson was able in 1919 to veto membership of Mexico and Costa Rica in the League of Nations. More fundamental is the objection previously noted that non-Members are not legally bound by obligations to settle their disputes by peaceful means, to refrain from the threat or use of force, or to assist in the maintenance of peace and security, even though the Organization assumes a police function in regard to non-Members. It is for this reason that I welcomed the admission of Argentina to the United Nations and deplore the exclusion of Spain, with or without Franco. The arguments used to oppose the admission of Argentina seem to me quite irrelevant. Fortunately for most of the states represented at San Francisco, the United Nations Charter contains no provisions relative to the "representative" or democratic character of its Members' governments. The United Nations is not democratic in structure or membership; nor is it fascist or communist: it is a functional security organization, and its tasks will presumably be lightened to the extent that states assume basic obligations for the preservation of peace.

For the same reason I deplore the provision in Article 6 of the Charter for the expulsion of Members who persistently violate the Principles of the Charter. Soviet Russia was expelled from the League of Nations; and history shows that expelled states continue to exist and to exert a potent influence in international affairs. The attempt to punish a state by releasing it from fundamentally necessary

146

international obligations is worse than ironical; it is a confession of failure. On the other hand, the idea contained in Article 5 of the Charter of suspending a Member "against which preventive or enforcement action has been taken by the Security Council" is excellent in principle, since the Member is suspended only "from the exercise of the rights and privileges of membership," and is not released from the obligations of membership. The procedure is faulty in granting to the Big Five a veto on suspension or restoration of rights and, possibly, in failing to provide for suspension on other grounds.

Unlike the Covenant of the League of Nations, the United Nations Charter contains no stipulation granting Members a legal right to withdraw from the Organization. However, apparently because of the failure to modify the rigidity of the Big Five veto on amendments to the Charter, the Report of the First Commission, approved by the San Francisco Conference in its ninth Plenary Session, contained a Declaration on Withdrawal, reading in part as follows: "The Committee adopts the view that the Charter should not make express provision either to permit or to prohibit withdrawal from the Organization. . . . If, however, a Member because of exceptional circumstances feels constrained to withdraw, and leave the burden of maintaining international peace and security on the other Members, it is not the purpose of the Organization to compel that Member to continue its cooperation in the Organization." To this statement the Soviet delegation took violent exception, not because it seemed to sanction a right of withdrawal, but because, said its delegate, "such right is an expression of state sovereignty and should not be reviled in advance." If the exercise of this alleged right of withdrawal were as unlikely as the example cited by Mr. Gromyko, we would probably have little to fear. He pointed to Article 17 of the Soviet Constitution which reads: "To every Union Republic is reserved the right freely to secede from the Union of Soviet Socialist Republics," and proudly referred to it as "this right of sovereign states" and "a most striking manifestation of democracy." Unfortunately, however, the Declaration on Withdrawal continues: "Nor would a Member be bound to remain in the Organization" if its rights and obligations as a Member were modified by an amendment which it found unacceptable, or if a proposed amendment, duly accepted by a two-thirds majority, failed to secure the necessary ratifications.

These concessions to the sovereign equality of states are not in the Charter and are contrary to the spirit of Article 108 of the Charter, which reads: "Amendments to the present Charter shall come into force for all Members of the United Nations when they have been adopted by a vote of two thirds of the members of the General Assembly and ratified in accordance with their respective constitutional processes by two thirds of the Members of the United Nations, including all the permanent members of the Security Council." Since the Charter contains no such proviso as is found in the League Covenant, namely, that "no such amendments shall bind any Member of the League which signifies its dissent therefrom, but in that case it shall cease to be a Member of the League"; and since, in the absence of a permissive treaty provision, no state may lawfully cease to be a party to an inconvenient

148

treaty, it is my opinion that at most the alleged Declaration on Withdrawal sanctions only nonparticipation in the affairs of the United Nations, and that the nonparticipating state would still be a Member and would continue legally subject to the obligations of the Charter.[2]

All Members of the United Nations are members of the General Assembly, each state having one vote. This recognition of the legal equality of states is not lessened by the fact that the unanimity rule which prevailed in the League of Nations Assembly is completely discarded in the United Nations General Assembly. Decisions on important questions—such as recommendations with respect to peace and security; the election of members to the Security Council, the Economic and Social Council, the Trusteeship Council; the admission, suspension, or expulsion of Members; and budgetary questions—are by a two-thirds majority of the members present and voting. Decisions on other questions are by simple majority of the states present and voting. This is a great advance over League of Nations procedure, but is compensated by the much reduced authority of the General Assembly as compared with the League Assembly. Where the latter had equal jurisdiction with the League of Nations Council to "deal at its meetings with any matter within the sphere of action of the League or affecting the peace of the world," the powers of the United Nations General Assembly are largely limited to discussion and recommendation. Its powers of discussion are broad enough to include any matter within the scope

---

[2] See, however, Secretary of State Stettinius' *Report to the President on the . . . San Francisco Conference, June 26, 1945.* Department of State *Conference Series* No. 71, pp. 47–49.

of the Charter, including consideration of general principles relating to the maintenance of peace and even specific questions relating to the maintenance of peace, but "any such question on which action is necessary" must be referred to the Security Council either before or after discussion. "The General Assembly may call the attention of the Security Council to situations which are likely to endanger international peace and security," but its power to make recommendations is strictly limited by the provisions of Article 12: "While the Security Council is exercising in respect of any dispute or situation the functions assigned to it in the present Charter, the General Assembly shall not make any recommendation with regard to that dispute or situation unless the Security Council so requests." Subject to the same exception, the General Assembly has the potentially important power to recommend measures for the peaceful adjustment of any situations likely to impair the general welfare, including situations resulting from violation of the Purposes and Principles of the Charter (Art. 14).

The General Assembly also has important budgetary powers, the power to receive and consider annual and special reports from the Security Council and other organs, and power to initiate studies and make recommendations for international co-operation on political, legal, and economic, social, cultural, educational, health, and human welfare matters and to further their development through the Economic and Social Council. Because of the sovereignty of states, explained former Secretary of State Stettinius to the Senate Foreign Relations Committee, the General Assembly does not have legislative power. However, through its agencies, it may prepare and consider draft conventions and call international conferences for their adoption, presumably subject to the old rule of ratification.

Turning now to the Security Council, we find that "primary responsibility for the maintenance of international peace and security" is conferred (Art. 24) on the Security Council by the Members of the United Nations, who "agree that in carrying out its duties under this responsibility the Security Council acts on their behalf," in accordance with the Purposes and Principles of the United Nations. And Article 25 contains one of the most important stipulations in the Charter: "The Members of the United Nations agree to accept and carry out the decisions of the Security Council in accordance with the present Charter." The importance of this obligation with reference to concepts of sovereignty and equality can be seen at a glance, but its full significance appears only from an examination of the powers of the Security Council. Although the major responsibility for the maintenance and enforcement of peace is placed on the Security Council, all Members of the United Nations are obligated to settle their international disputes by peaceful means of their own choice (Arts. 2 [3], 33 [1]). The Security Council may "call upon" the parties to settle their disputes by such means (Art. 33 [2]); it may investigate any dispute, or any situation which might lead to international friction or give rise to a dispute, in order to determine whether its continuance is likely to endanger peace or security (Art. 34); and it may at any stage recommend appropriate procedures or methods of adjustment of such disputes or situations (Art. 36).

If the parties to such a dispute fail to settle it, they are obligated to refer it to the Security Council (Art. 37 [1]), which may decide either to "recommend" methods of settlement or terms of settlement (Art. 37 [2]). It should be noted here that no party to a dispute has a vote in decisions of the Security Council for the pacific settlement of disputes (Art. 27 [3]).

Of prime importance is the power of the Security Council to determine the existence of occasions for the application of sanctions or enforcement measures to maintain or restore peace (Art. 39). Under the League of Nations Covenant "resort to war" was the only occasion for employing sanctions, and under League practice each Member decided for itself whether the occasion existed and whether to employ sanctions. In the United Nations Charter, the occasions for employing sanctions are either a threat to the peace or a breach of the peace, and the existence of these conditions is determined by the Security Council in a decision legally binding on all Members of the United Nations. Furthermore, the Security Council may decide what diplomatic or economic sanctions are to be employed and "it may call upon the Members of the United Nations to apply such measures" (Art. 41). If the Security Council decides that economic sanctions are inadequate, "it may take such action by air, sea, or land forces as may be necessary to maintain or restore international peace and security" (Art. 42). Since the United Nations will have no international army of its own, all Members of the United Nations obligate themselves "to make available to the Security Council, on its call and in accordance with a special agreement or agreements, armed forces, assistance, and facilities" necessary for maintaining peace and security (Art. 43). These agreements are to be negotiated on the initiative of the Security Council and are subject to ratification by signatory states. Within the limits of these special agreements, Member states will be obligated to "hold immediately available" national air-force contingents to enable the United Nations to take urgent military measures (Art. 45).

In a last-minute compromise, the San Francisco Conference introduced a proviso that when the Security Council "has decided to use force" a state not a member of the Security Council has a right to participate in the decisions of the Security Council concerning the employment of its armed forces (Art. 44), although the voting provisions of the Security Council give that state no veto over the fulfillment of its obligations (Art. 27 [3]); and Article 48 states that "the action required to carry out the decisions of the Security Council for the maintenance of international peace and security shall be taken by all the Members of the United Nations or by some of them, as the Security Council may determine."

A weasel-worded chapter (XVII) which has received little public attention is entitled "Transitional Security Arrangements." By it the states signing the Charter agree that, pending the coming into force of such special military agreements as in the opinion of the Security Council enable it to take military action in accordance with the Charter, the Big Five will assume responsibility for joint action for maintaining peace (Art. 106); and "nothing in the present Charter shall invalidate or preclude action . . . taken or authorized" in relation to an enemy state as a result of the present war "by Governments having responsibility for such action." The principle of delegated responsibility, at least, is clear; and, in time, we may learn whether it is intended that a new Conference of Ambassadors, Paris style, is to operate alongside the Security Council.[3]

It is not the purpose of this paper to discuss all the interesting and important features of the United Nations Organization; the Economic and Social Council, the International Trusteeship System, and the International Court of Justice are worthy of detailed consideration on their own merits. It has been my purpose to place the San Francisco Charter in its conceptual and historical setting and to indicate its approach to the problem of national security in a world of power. My concluding reflections can be expressed within the same framework of ideas.

In terms of sovereignty, equality, and governmental function, the Charter marks an important advance in some matters and, in others, no progress at all. The Charter will come into force when it has been ratified by each of the Big Five and a majority of the other signatories, but only for the states which ratify. The traditional rule that no state is bound without its consent is also perpetuated in the failure of the Charter to grant to any organ the legislative power to enact binding rules of international law, and in the failure of the Statute of the International Court of Justice to confer compulsory jurisdiction on the Court.

---

[3] Since the above was written, the Tripartite Conference of Berlin (Potsdam) agreed to the establishment of a Council of Foreign Ministers representing the United Kingdom, the Union of Soviet Socialist Republics, China, France, and the United States. See Department of State *Bulletin*, Vol. XIII, No. 319 (August 5, 1945), p. 153. The first meeting of the Council was held in London in September, 1945.

On the other hand, broad jurisdictional powers and authority to make crucially important decisions, legally binding on all Members of the United Nations, are delegated to the Security Council. Moreover, in the delegation of this authority, concepts of sovereignty and equality have been sharply tempered for all Members of the United Nations except the Great Powers. Indeed, throughout the Charter, the unanimity rule—so prized by all states at Geneva— has been struck a series of body blows. In the Economic and Social Council and the Trusteeship Council, all decisions are by simple majority; in the General Assembly, important decisions are by a two-thirds vote and all others by a majority; in the Security Council, decisions are ordinarily by a majority of seven out of eleven, with the important proviso that the affirmative vote of seven must include the concurring votes of the five permanent members on most matters. Within the Organization, the unanimity rule thus becomes the perquisite of Great Powers. The point was not yielded without great travail but the common-sense view prevailed that a functioning security organization would be a better guaranty of security than a somewhat hollow claim to juridical equality. The renunciation by forty-five states of the veto on executive action for the maintenance of peace is a noteworthy advance in international government, and, in time, the development of adequate legislative and judicial authority in the international sphere may follow the same pattern of development.

In terms of power, the form given to the United Nations Organization is grounded realistically on a fact and a proposal: the fact is the preponderant power of the British Empire, the Soviet Union, and the United States to maintain peace in the postwar world, and the proposal is that major responsibility for enforcing peace should be placed in the hands of those who have the power. The proposal came from the three World Powers themselves. Conceivably, they could have sought world peace without any organization, each Power stabilizing great areas and relying on its own military and industrial might to curb the other two. That they preferred an international security organization, which, while recognizing their preponderant power to maintain peace, places legal restraints on the use of force by all states—themselves included—is an expression of their belief that by no other means could a state provide so well for itself against all possible combinations of power.

Yet it is on the point of collective security that the Charter is so seriously deficient. Not only can any one of the five permanent members of the Security Council veto enforcement action against itself, but it can veto any action—whether preventive or punitive—in disputes to which it is not a party. The most that can be said for this veto is that the assumption of the major responsibility for enforcing peace by those states with the necessary power, inevitably involves a recognition of their special interest in the determination of when and how they shall fulfill this obligation; and that an irresponsible use of the veto would be so flagrant a violation of the Purposes and Principles of the Charter that, to paraphrase Senator Vandenberg, a state which "breaks this contract will stand in naked infamy before the embattled conscience of an outraged world."

This rhetoric should not blind us to the fact that there are no means by which the smaller states can enforce the

INTERNATIONAL ORGANIZATION

delegated responsibility for peace. In this, however, they are no worse off than before. We must remember that the small states delegate responsibility and legal authority, not power; and the states which possess great power set forth the terms on which they will accept responsibility for its use in the common good. In their ability to collaborate in the performance of this heavy responsibility lies the key to the postwar years, for protection against violence is essential if men are to accomplish, in the words of the Charter, the promotion of social progress and better standards of life in larger freedom.

EDUCATING
AMERICAN CITIZENS

*

BY GEORGE D. STODDARD

*COMMISSIONER OF EDUCATION
OF THE STATE OF NEW YORK*

IN THIS LECTURE I propose to examine American education from two points of view: first, a description of its structure, and, second, an analysis of events and programs within that structure.

I

The education of the American child has begun traditionally at the year six. At that time the average child is mature enough to profit by group instruction in reading, writing, and arithmetic. Since the cardinal aims of education had long included other qualities of child development, such as physical fitness, knowledge of the world, and character, it became clear that education could be supported well below the age of six. Accordingly, the kindergarten movement began. It has flourished moderately in this country, with perhaps 30 per cent of the children enjoying its privileges.

Next, a battery of researches in child behavior revealed what was already known to the close observer: children learn at all ages, and in the preschool years learning is significant and dramatic. The beginnings of language are not in reading but in talking; the beginnings of science are not in the classroom but in the child's early discovery, often painful, of cause and effect relations. The preschool child is sensitive to art and human affairs. We discovered some years ago at the University of Iowa that he is aware of pulchritude in the teaching staff. The young child learns about people by close contact with such complex items as parents and older siblings. He discovers how to get what he wants. If the parents are naïve, selfish, or malicious at this time of the child's life, he will nail down for later use some nasty ways of dealing with other persons. In short, confirmed bad behavior at this early level may be a forerunner of malnutrition, instability, or mental deficiency.

I attach these big words to little children, not to alarm parents already aroused over their failure to understand and control children, but in order that we may not regard the downward extension of education as a casual reform. A good kindergarten is as useful as the fourth grade, a good nursery school as defensible as a college.

In the State of New York the recent statute revising the formulas for state aid gives encouragement to these lower years. For the first time school boards will receive full state aid for the attendance of kindergarten children, attendance being counted down to the chronological age of four. In New York the kindergarten child is no longer a thing apart. We may expect to develop widely over the State a two-year kindergarten plan in which the second year is a genuine advance in mental and social expectation over the first. Both years should contribute notably to the preparedness of a child to enter upon the regular work of the primary grades.

It is not possible in a brief paper to review the arguments and researches that support early childhood education. The obvious questions have been asked and answered; we

162

have a thousand research studies and ten thousand testimonials. We know why it is good to remove children from their homes for a certain portion of the day. We understand the value to children of social contact outside the home; relief of the mother from their constant supervision appears to be refreshing to all concerned.

We have learned that cost accounting in educational matters must be connected with results: a slight reduction in such afflictions as ill health, physical impairment, feeblemindedness, delinquency, or abnormality would pay the bill for public education from nursery school to university. The total cost of crime alone is five times the annual bill for public education in the United States.

Within the structure of education from grade one to the end of high school, a major reform in the State of New York is the full recognition, in the law of 1945, of grades seven to twelve as secondary education. The seventh grade pupil, wherever he may be found, is recognized as a high school pupil; state aid is made available accordingly. Eventually this should mean the elimination of the old type grades seven and eight; it should lead to improvement within the junior and senior high school programs.

At these levels the child is growing up. Toward the end of high school he really becomes a man. He lacks experience, but who does not? If ever, he is then eager, healthy, and bright. There are denied to him at the age of eighteen only a few political and social rights, and our justification for such denial on the political side is steadily weakening. With all his defects, the high school graduate of the mid-twentieth century is a fine example of democratic education applied on a wide scale. Graduates of our better schools are well prepared in civic understanding. Youth everywhere are prepared to fight and die for their country. This we witness without a proper realization of its significance in the awarding of peacetime benefits.

However, it is not my purpose to say much about the structure of education at traditional levels. The high school is the glory of the American educational system; it is, in fact, the envy of the whole world. As one country after another discovers its potentialities, it will be copied, extended, and enriched. The chief wealth of a nation is to be found in its natural resources, its cultural heritage, and its youth population. In America we are rich in all three, but we have been careless.

For example, in some states only a small portion of the youth population finishes high school. In the State of New York the picture is not gratifying: of all the young people who enter the seventh grade, only 50 per cent are graduated from high school; of these, only 40 per cent take any advanced work whatever. We find no defense for this in terms of the measured mental ability of youth. In some states the percentage of high school graduation is over seventy, and it is not clear that their young people are more talented or their schools of lower standard. Rather we should look for such items as these:

(1) Differences in cultural expectation for the poor, the colored, the alien, and the children of foreign parentage;
(2) Geographical differences within a state, for example, as between rural and urban; and
(3) A rigidity in curriculum, such that the pulling power of further education is reduced. (Thus, if a curriculum is pointed mainly toward college preparation, it appeals to the 20 per cent of youth who undertake some kind of college work.)

These observations lead directly to a consideration of the Regents' Plan for Postwar Education in the State of New York, with special reference to the development of a system of Institutes of Applied Arts and Sciences. To quote:

The Regents recognize . . . that there are still large numbers of boys and girls who do not take advantage of the full program of secondary education, and, further, that many high school graduates will expect to apply their knowledge in a vocation on completing an educational program two years beyond high school graduation. Under proper guidance, a high school graduate should be able to clarify his principal vocational interests, if he does not plan to spend four years in college. At the same time, every encouragement should be given to the boy or girl entering upon a four-year college curriculum.

The Regents' solution to this problem is three-fold:

1) The development of a new system of state-supported Institutes of Applied Arts and Sciences

2) The establishment of a new system of scholarships, and

3) The strengthening of offerings in the present system of state institutes and state colleges.[1]

It is expected that the curriculums of the Institutes will include:

1) A basic preparation for selected arts, technologies and sub-professions which require a technical proficiency not reached in high school programs. Some of the indicated occupations are those of: draftsmen, electrical technicians, store operators, dietitians, radio technicians, workers in hospitals, and in building, automotive, aviation and photographic services, laboratories, graphic arts, transportation, communication and electronics.

---

[1] *Regents Plan for Postwar Education in the State of New York* (Albany, The University of the State of New York, The State Education Department, 1944), p. 12.

165

2) Related offerings in arts and sciences.

3) Personal and civic arts designed to further the general welfare and understanding of the students. Instruction in English, social science and other liberal subjects is considered essential to personal growth and citizenship.[2]

This great plan for the State of New York was endorsed by Governor Thomas E. Dewey in his annual message and was referred to a temporary commission in a 1945 statute. Among other powers, the Commission is authorized to "determine the location of state institutes on a regional basis including state institutes in New York city, and authorize the state education department to accept sites, acquire property and facilities, lease buildings and to take such other steps as may be necessary to establish and develop such institutes." The Commission, under the chairmanship of Senator Benjamin F. Feinberg, is now at work developing the institute program.

It may be said that support for such a plan has been widespread over the country. For example, Mr. Clement C. Williams, a distinguished engineer, formerly President of Lehigh University, in an article entitled "Tertiary Technical Education a Postwar Need" makes this recommendation with reference to the United States as a whole:

To revise the general scheme of education to accord more closely with the diversity in youth aptitudes, to align training with the occupational patterns, and to preserve the standards of higher education, the terminal phases of the junior-college plan should be extended to include a system of technical institutes for training elastically up to two years following high school. Incidentally, the schooling for refurbishing the war veterans would largely fall within this provision. Our present educational system prepares about 80 per cent for white-collar

[2] *Ibid.*, p. 14.

jobs while the occupational pattern is nearer 20 per cent white collar and 80 per cent overalls. The resulting crowds flocking to universities endanger standards as compared to universities in older civilizations, and dissipate resources and scholarly talent. . . .

Therefore postwar planning bodies will be well advised to give further consideration to terminal tertiary technical training with a view to efficiency and economy. In order to effect a nicer adjustment to the aptitudes of youth and to the variety in occupational patterns, technological education should be broad and flexible and not limited to the traditional model of the baccalaureate.[3]

In other areas of education the State of New York is developing pioneering projects. The recently established New York State School of Industrial and Labor Relations at this university is an outstanding example. Having sat through the discussions of the Board of Temporary Trustees of the Institute and having listened to the informed and at times impassioned pleas of leaders in government, labor, and agriculture, I share with President Edmund Ezra Day the discovery that education has powerful friends. In fact, a composite record of the extemporaneous addresses of Mr. Irving M. Ives, Mr. Louis Hollander, and Mr. Howard E. Babcock during the preschool days of the Institute would make a suitable archive to be placed in its cornerstone.

On other fronts additions to the educational structure are more difficult to classify. The new law for veterans' scholarships in the State offers 2,400 grants at $350 a year, which is the average cost of tuition in New York colleges. Since the older scholarships are retained at $100, the veterans' plan has the effect of bringing more closely within

---

[3] *School and Society,* LXII (July 21, 1945), 37.

a general structure of educational opportunity the offerings of the numerous colleges. We are hopeful that within the year this opportunity may be extended widely to competent youth just graduating from high school. The theory underlying the development of a new scholarship program for New York rests simply upon the full utilization of public and private facilities in higher education. Public facilities in higher education, except in the city of New York, are lacking in fields other than agriculture, home economics, veterinary surgery, ceramics, teacher training, maritime service, and forestry. The effect of strengthening the scholarship program will be to offer comparable financial aid to students whose interests lie in the liberal arts and sciences. In fact, strong theoretical and practical support could be mobilized for a system of scholarships extending through graduate and professional work for all students of outstanding ability.

Increasingly the State gives aid to special programs. The work of various departments in preventing juvenile delinquency is illustrative. These programs are centered in a Youth Service Commission whose major functions have been outlined in a recent report:

1. Coordinate related activities and efforts of State departments working in the field of delinquency.
2. Stimulate localities to set up programs for coordinating the total community program.
3. Assist schools to extend their particular contribution in locating and helping vulnerables.
4. Assist localities to extend recreational programs so as to broaden the content and to reach all children.
5. Assist in extending the child care program so as to reach all homes needing such help.

6. Assist in recruiting and training leaders 'for voluntary youth organizations.
7. Assist localities to secure needed specialized services such as psychiatric, psychological and social work services when existing agencies are not able to supply them.
8. Assist localities in making surveys of existing needs and available resources.
9. Assist in appraising the achievement of local programs.
10. Serve in a general consultive capacity—acting as a clearing house, developing materials, arranging conferences.
11. Develop and maintain enlightened public opinion in support of a program to prevent delinquency.
12. Formulating a pattern to provide for this work on a continuing basis.[4]

Through grants in aid to agencies and communities, it should be possible to develop a better understanding of delinquency and to set up structures within which vulnerable children and youth may be located and "corrected." I have placed the word "corrected" in quotes, for everybody knows that delinquency, more often than not, implies a fundamental breakdown in the home and related social agencies.

Nobody wants delinquency, but almost everybody contributes to it. The failure lies in our ignorance of the deep-seated drives of children and youth. We fail properly to feed, house, guide, and educate children and then worry about the defective outcome. For millions of parents over the country, such concepts as self-reliance, mental hygiene, recreation, social development, and emotional outlet are just big words used by persons who have nothing better to do with their time. Every psychologist, sociologist, and

---

[4] *Preventing Juvenile Delinquency* (Interim Report of the Interdepartmental Committee on Delinquency, Part I; Albany, 1944), pp. 11–12.

judge who has looked into the problem of delinquency discovers within its ramifying context an indictment of common patterns of control and guidance. The child is not naturally rebellious or delinquent, but he is naturally a bundle of energy and a personality. He is a social creature in need of security, friendship, and love. It is amazing how scarce these articles become, even in a democracy.

Nobody believes that statutes, commissions, youth bureaus, and community councils supply the full answer. We believe, however, that they may bring out the chief issues in such a fashion as to relate cause and effect, thus encouraging a new mobilization of knowledge and good will. We shall try to break away from general apathy, wishful thinking, and the irrational effects of remorse.

Again, the legislature of 1945 has added to our structure by granting substantial state aid for summer schools, night high schools, and part-time continuation schools, together with an allowance for classes in adult education. All these programs have the effect of making more flexible, not only the calendar of educational offerings, but their character. For example, adults as pupils are both more and less than children. They know more, but in certain areas they may have developed a resistance to learning. They are free to come or stay away—a disturbing factor for the poor teacher.

What adults may study under a combination of local and state support is, for the present year, a matter of experimentation. We are not willing to admit that evenings devoted to pinochle, ping-pong, or canoeing fall within the scope of adult education, delightful as they may be. At the same time we abhor the doctrine that learning takes place only in an atmosphere of frustration or dislike. Some recreational programs, such as those found within the regular offerings of schools and colleges, may be said to be within bounds. The hope is that, as they return for schooling appropriate to their maturity and individuality, adults will seek the more serious purposes of life. Along intellectual lines, everybody is short on something. A new interest in science or in the dramatic, graphic, and literary arts might well serve as a fountainhead.

Persons who abandoned high school or college on grounds that were not valid or that have been cut away over the years should undertake new learning. High school students who for patriotic or financial reasons left their classrooms should now return under proper auspices.

No high school should ever cross a student off its rolls except under one of two conditions: (1) the student has graduated; or (2) the student, by reason of measured limitations, should leave the academic life in order to develop suitable occupational skills on the job. In high school, institute, or college, anybody who is designated "ex," followed by a calendar year, is unfinished and unrequited. Programs are needed that will re-enlist his attention and enthusiasm. It will be no minor matter if such programs quietly lead to the appropriate high school diploma or college degree. In short, I am recommending that we tap the largest alumni body in the world—the endless scroll of those who began something but, frequently through no fault of their own, failed to carry it through. It is trite to say that it is never too late to learn, but it is not customary to provide the implementation.

In rural education the State of New York is carrying

through its extensive program of centralization. The central rural school system, only twenty years old, numbers 321 districts. They have replaced over 4,700 separate small districts that could not offer full advantages to any child. There are still several thousand small districts, but it is encouraging to report that a great many central districts are in process of formation. Following the war these plans should quickly become effective. It is well known that some of the older central districts are too small. Perhaps insufficient attention was paid to trends in transportation. Researches are under way in the Education Department, in collaboration with leaders in education and rural development over the State, that should indicate improvements within the central plan. We may come to a clustering of districts for certain supervisory and guidance services.

Similarly, the work in agriculture at the high school, institute, and college level is being studied by a committee on rural educational services that includes Regents, members of the Cornell Board of Trustees, and a joint staff. The aim is to develop a completely integrated pattern of rural educational services, bringing all children and youth to a high level of recognition and opportunity.

II

Thus far I have talked about the structural aspects particularly in the State of ·New York, with only occasional reference to functions and goals. It will be well, therefore, to look more closely into the programs which education, however organized, is expected to develop.

It should be said at once that, in every type of education

and at every level, pupil progress is a derivative of teacher excellence. Broadly speaking, good teaching *is* good education. The good teacher understands the motives of pupils and has access to effective methods and materials. The good teacher, moreover, has something to say about curriculum development; he is not simply a purveyor of goods on the shelf. Truly the long and rewarding campaign for better teaching embraces the whole range of the educative process.

Why, for example, do we want safe buildings, well heated, lighted, and ventilated? The answer is, for the purpose of good instruction. Where lighting or ventilation is bad, fatigue sets in, and fatigue by definition is a bodily process opposed to achievement. Similarly, in the one- or two-room school it is not feasible to arrange classes in areas of specialization or in terms of the needs of children. The only special argument for tenure and good salary schedules is in terms of their effect upon the development of children. The concept of the good teacher carries with it knowledge, character, and stability—all favorable to learning and adjustment on the part of the pupil.

The school, as an extension of the home, calls for excellence in learning, behavior, and attitude beyond anything required in an ordinary job. It is noteworthy, however, that business enterprises give increasing attention to optimum conditions for nutrition, fresh air, and harmonious surrounding. Subtle factors in morale are no longer beyond the interest of managers seeking the maximum in production.

As we go above these home management standards, we discover, for example, that legislatures regularly mandate the schools to take positive measures toward health and physical fitness.

In the State of New York these measures are outlined in various parts of the Education Law. In spite of the detail and the attention paid to health instruction, exercises, and sports, the record of the schools is not always satisfactory. It is true that since the early days of the war some excellent programs have been established, and we know that certain high schools have achieved a remarkable record in correcting physical defects.

However, statistics on military induction are rather depressing; as we get beyond the teen ages we discover an appalling accumulation of ills and disabilities that do not speak well for our national health. Accordingly, we have been trying to improve the school programs. In spite of the crowded condition of the so-called constants or mandates in the high school program, the Regents have added an additional required subject, namely, health instruction, with the understanding that arrangements can be flexible as to schedule. Nobody believes that a measure of this kind by itself alone will be effective, but it is a step in the right direction.

We can teach health and improve habits, but some of the conditions do not fall within the school program. It can be demonstrated, for example, that health is related to these factors:

(1) Hereditary conditions
(2) Nutrition, especially in early childhood
(3) Medical diagnosis and treatment
(4) Exercise and recreation
(5) Emotional attitudes, and
(6) Information and health habits.

174

Now if you can imagine yourself a teacher in the ninth grade, let us ask concretely what you could do about such factors. Is it not true that most of the responsibility assigned to the classroom teacher is contained in the last three items? You could teach about microbes and safety precautions, and with good effect, too. Children, like dogs, used to drink out of puddles and streams, especially if the water was clear. It would be an uncommon child now who would take such a risk. Also most children coming up through school have a fair knowledge of first aid and safety. I have observed the careful way in which most children cross a busy intersection, although I am not much impressed with their cautiousness at the wheel of a car a few years later. All these skills and habit patterns are of one piece; schools that teach personal cleanliness and pure food may go on, as some high schools and colleges have done, to teach the habit patterns of skillful driving. In this State good driving is being taught as a war measure.

We are undertaking a survey which may lead to a revision of the statute and of school practice in the field of health and physical fitness. We are anxious to get away from omnibus solutions in which the intentions are good but the effects difficult to measure.

We believe that in health and physical education, as in everything else, we cannot get something for nothing. Good health starts even before birth—as some wit has remarked, in the proper selection of your parents. Then attention must be given to infancy, an important period but not a function of the school except as it may come within adult education. For the most part the guidance, feeding, and medical attention of infant and preschool child enter

175

into educational offerings on an informal basis. Increasingly, I believe, they will be considered a part of public instruction under the heading "parent education" or "adult education." The other factors listed above are similarly marginal with respect to the school's major responsibility. Medicine is primarily a private concern between the doctor and the patient, although public health, hospital, and compensation plans are increasingly important.

It is easy to forget, too, that the school year is short; it is only about one-half of the calendar year. Moreover, the average program, five or six hours in length, consumes less than one-half of the child's waking time. It is in the non-school periods that many matters related to health, physique, and emotional stability receive their definitive direction. Thus a child in school for half a day through half the year may be subjected to unfavorable pressures in the home and neighborhood. The best results in school are obtained when we have full co-operation and understanding among the school, the home, and the community. Good neighborhoods tend to produce good schools, and the schools in turn enrich the community life.

The key concept throughout the compulsory period of education is *general education* or *basic education*. It is true that this concept in the last year or two of the vocational high school is bent to fit immediate occupational demands. However, what I shall say on this subject should apply, if it applies at all, to a very substantial proportion of the pupils of the State, public and private.

"Knowledge indiscriminately gathered in for its own sake has no place in our concept of intelligence-at-work,

176

nor is it, according to Einstein, to be highly regarded as a preparation for life:

I want to oppose the idea that the school has to teach directly that special knowledge and those accomplishments which one has to use later directly in life. The demands of life are much too manifold to let such a specialized training in school appear possible. Apart from that, it seems to me, moreover, objectionable to treat the individual like a dead tool. The school should always have as its aim that the young man leave it as a harmonious personality, not as a specialist. This in my opinion is true in a certain sense even for technical schools, whose students will devote themselves to a quite definite profession. The development of general ability for independent thinking and judgment should always be placed foremost, not the acquisition of special knowledge. If a person masters the fundamentals of his subject and has learned to think and work independently, he will surely find his way and besides will better be able to adapt himself to progress and changes than the person whose training principally consists in the acquiring of detailed knowledge.[5]

". . . Let us assume that every effort has been made to provide superior school experiences, taking into account the full resources of home and community. Whatever the curriculum and the motivation, however high the standards of accomplishment, we shall discover fundamental differences in human ability and progress. This hypothetical school system would attempt to secure not uniformity, but a common denominator of learning, behavior, and social adequacy, beyond which pupils would be granted the privilege of wide variability, both outward and upward.

---

[5] "Some Thoughts Concerning Education," in *Proceedings of the Seventy-second Convocation of the University of the State of New York* (University of the State of New York Bulletin, 1936, No. 1100), p. 47.

The upward extension would be under the guidance of the finest teachers, who would devote to it the exquisite care of a Mill or a Montaigne." [6]

We should all admit that this hierarchal system has not as yet characterized most of our teaching or testing methods. At any grade level we tend to spread out in all directions, and even when we move upward we tend to move in the direction of greater feats of memory. Nevertheless, the whole system of grade classification and promotion implies a hierarchy of adjustment and accomplishment; twelfth grade English should be definitely in advance of eighth grade and not simply an old set of experiences under a new teacher at the same level of abstraction. Conceived in this fashion, it cannot be said that the high school curriculum is too long even for the most able students.

From the beginning, questions of ethics and morality have been paramount within the structure and function of the American school. Recently, when asked what we were doing to prevent discrimination in the schools of New York, we were taken aback by the nature of the question. The public schools work the other way around; they accept everybody on an equal basis. They are the friends and servants of all the people. The Constitution of the State, long before the present legislation on antidiscrimination, established a basis for social and civic equality within the schools. Our textbooks, syllabi, and teaching methods, with minor lapses, have sustained this mandate of the people. Curiously the so-called fairness of textbooks has resulted in some criticism: a well-known textbook was discovered

[6] George D. Stoddard, *The Meaning of Intelligence* (New York: The Macmillan Company, 1943), pp. 432–433.

to be friendly to foreign countries and other races, including the Japanese. Generally we have not taught children to hate anybody under any circumstances. When the Army received the products of our high schools and colleges, it felt called upon to introduce a period of indoctrination in order to induce good hating.

These episodes reveal the difficulties in a policy of tolerance. When does tolerance shade over into appeasement? No decent person could tolerate Hitler or Mussolini; they were against everything good in our civilization. We must teach discrimination to pupils, but relate it to a selection of what is good and what is bad. However, contempt and hatred rarely need be stirred up, for negative reactions are well entrenched. A positive emphasis on the good emotions will serve to counterbalance an ever-present capacity for anger, cruelty, and aggressiveness. In this respect the home, the school, and the church have much in common, arriving at similar goals by different routes. The school, above all others, stresses actual conduct within a social setting, paying less attention than the home to authority parental in type and less attention than the church to ideological substrata.

What we seek is a blending of scientific, humanistic, and spiritual forces in the education of the American citizen. In an age of science and technology he must be technically informed. In addition, he needs a rewarding artistic participation that may or may not be related to his working hours. He must understand that schooling is a part of life —a very precious part—and that it may be extended upward so long as personal growth is maintained.

I have implied that, as educators, we can weld the liberal

179

and the practical into a seamless whole. In closing may I return to this point?

Consider, for example, the returning veteran. He risked his life, and not for any simple reason. He cannot tell us all that happened and we could not understand. No matter what the casual remark, a man's life is not easily offered up. Let us not misunderstand these brave men; let us offer them the best, as age-long experience has defined it: a life of growth and achievement; a mixture of security and adventure; a place of understanding and affection; surroundings that combine efficiency and beauty. All this means planning and achievement on a vast scale. Let us carry this achievement to the familiar scene—to farm, village, city, and factory.

Behind these tangibles, but related through the compatibility of nature, will be found analogous creations in the social structure. Man can find within himself a great peace if he will courageously support his own best efforts. With the coming of world peace, growing out of man's latest invention—the United Nations—there will still be room for physical, ethical, and spiritual adventure. The search for the weapons of peace and the maintenance of a world structure for this alone will become a part of every teaching program. To fail in this lesson is to lose the benefit of all the others.

War never was a frontier, but a return to the primitive. Not war, but the bitter struggle against it, will survive to test us all.